CSM-EE

CLASSIFICATION AND STATISTICAL MANUAL OF EXTRASENSORY EXPERIENCES

CSM-EE

CLASSIFICATION AND STATISTICAL MANUAL OF EXTRASENSORY EXPERIENCES

THERESA M. KELLY, MsD.

Digital 1ˢᵗ Edition
Copyright © October 2014 by Theresa M. Kelly, MsD.
Center for Exceptional Human Experiences

For more publications by this author, please visit: http://qpsychics.com

ISBN 978-1-312-62493-1

Contents

TELEPATHIC EXPERIENCES

EE1 or EE(GT)   **GENERAL TELEPATHY**

CRITERIA

1.   Mind-to-mind communication.

2.   Involves two or more individuals.

3.   The agent and subject, percipient, or all participants, are living organisms.

SEVERITY

Stable/Functional, Mild, Moderate, Severe.

EE1.1 or EE(TC)   **TELEPATHIC COGNITIVE EXPERIENCES**

CRITERIA

1.   Information is received by the percipient.

2.   Information received is in first person perspective (e.g. If visual: the image received is from the subject's perspective), or narrative (e.g. If auditory: the words received are from the subject's perspective "He/She is happy today;" or occasionally '"I am happy today;" with additional information in reference to the subject).

3.   Subconscious need for information acquisition present at the time of the experience.

SPECIFIERS

Spontaneous, Intentional; Adaptive, Decisive; Single Episode, Episodic, Continuous; Dream, Intuitive Impression/Emotional, Auditory H, Visual H, Tactile H, Somatic H,

Olfactory H, Gustatory H, Compound.

EE1.2 or EE(TI)   **TELEPATHIC INTERACTIVE EXPERIENCES**

CRITERIA

1. Information is sent by the agent.

2. Information is sent in second-person perspective (e.g. If visual: the image impressed is from the agents perspective) or narrative (e.g. If auditory: the words impressed are from the agents perspective i.e. "You want to behave this way.").

3. Subconscious need for information impression present at the time of the experience.

SPECIFIERS

Spontaneous, Intentional; Adaptive, Directive; Suggestive, Compulsive; Single Episode, Episodic, Continuous; Dream, Intuitive Impression/Emotional, Auditory H, Visual H, Tactile H, Somatic H, Olfactory H, Gustatory H, Compound.

EE1.3 or EE(TS)   **TELEPATHIC SIMULATIVE EXPERIENCES**

CRITERIA

1. Information is shared between the telepathist and one or more participants.

2. Information is shared in first-person plural perspective (e.g. If visual: the image is shared with the telepathist and all participants, and is from a group perspective involving all other participants) or narrative (e.g. If auditory: the words shared are in first-person plural perspective i.e. "We want to behave this way.").

3. Subconscious need for information sharing present at the time of the experience.

SPECIFIERS

Spontaneous, Intentional; Adaptive, Directive; Input, Output; Single Episode, Episodic, Continuous; Dream, Intuitive Impression/Emotional, Auditory H, Visual H, Tactile H, Somatic H, Olfactory H, Gustatory H, Compound.

CLAIRVOYANT EXPERIENCES

EE2 or EE(GC)   **GENERAL CLAIRVOYANCE**

CRITERIA

1. Mind-to-environment/nature (including information about a person), mind-to-object, or mind-to-entity communication/effect.
2. Involves one or more environments, objects, entities, or indirect (about) information pertaining to a person's situation.

3. The source of the information is not a living organism (i.e. the source is the environment/nature, a discarnate entity, or other entity), but information can be obtained, or probability influenced, pertaining to a living organism's situation (e.g. health, environment, current events).

SEVERITY

Stable/Functional, Mild, Moderate, Severe.

EE2.1 or EE(CC)   **CLAIRVOYANT COGNITIVE EXPERIENCES**

CRITERIA

1. Information is received by the percipient through mind-to-environment (including information about

2. a person) mind-to-object, or mind-to-entity communication.

3. Information received is in third person perspective (e.g. If visual: the image received is viewed as though the percipient is looking at an event, object, or looking at the individual within their surroundings (i.e. rather than looking through the eyes of an individual -- telepathy), or narrative (e.g. If auditory: the words received are from the sources' perspective "You will have a fortunate day," or "She misses you dearly."

4. Subconscious need for information acquisition present at the time of the experience.

SPECIFIERS

Spontaneous, Intentional; Adaptive, Decisive; Precognition, Contemporaneous, Retro/Postcognition; Nature, Discarnate, Other Entity; Individual, Object, Event; Single Episode, Episodic, Continuous; Dream, Intuitive Impression/Emotional, Auditory H, Visual H, Tactile H, Somatic H, Olfactory H, Gustatory H, Compound.

EE2.2 or EE(CI)   **CLAIRVOYANT INTERACTIVE EXPERIENCES**

CRITERIA

1. Information is conveyed through the experient via some mode of entity interaction/occupation.

2. Information is conveyed in first person narrative (i.e. the entity is talking through the experient and refers to them self (the entity) as "I").

3. Subconscious need for information conveyance present at the time of the experience.

SPECIFIERS

Spontaneous, Intentional; Adaptive, Decisive; Trance Mediumship, Channeling; Nature, Discarnate, Other Entity; Individual, Object, Event; Single Episode, Episodic, Continuous; Automatism, Xenoglossy, Physical Mediumship, Psychopompic Activity, Compound.

EE2.3 or EE(CS) **CLAIRVOYANT SIMULATIVE EXPERIENCES**

CRITERIA (INPUT)

1. Accommodating information is shared between the experient and the environment/Nature.

2. Accommodating information is shared with seemingly no perspective other than the self (e.g. if visual: the image is shown from the experient's perspective, if auditory: information is heard in-mind in the experient's own voice and is heard in first-person narrative such as "I think," or "I feel."

3. Subconscious need for accommodating information present at the time of the experience.

CRITERIA (OUTPUT)

1. An accommodating and meaningful coincidence, synchronistic event, or probability shift has occurred.

2. Subconscious need for an accommodating meaningful coincidence, synchronistic event, or probability shift present at the time of the experience.

SPECIFIERS

Spontaneous, Intentional; Adaptive, Directive; Input, Output; Individual, Place, Object, Idea, Event; Precognition, Probability Shifting, Contemporaneous, Retro/Postcognition, Historical Shifting; Single Episode, Episodic, Continuous; Dream, Intuitive Impression/Emotional, Auditory, Visual, Other, Compound.

EMPATHIC EXPERIENCES

EE3 or EE(GE)   **GENERAL EMPATHY**

CRITERIA

1. Mind-to-Mind, or mind-to-environment, emotional communication.
2. Involves one or more individuals, or involves one or more environments and indirect emotional information pertaining to one or more groups of individuals.
3. The subject or participant is a living organism (e.g. human, animal), or the target group is comprised of living organisms (e.g. human, animal) and the information obtained about them is in reference to the emotional state of the target group (e.g. emotions towards community or national health, politics, current events, etc.).

SEVERITY

Stable/Functional, Mild, Moderate, Severe.

EE3.1 or EE(EC)   **EMPATHIC COGNITIVE EXPERIENCES**

CRITERIA

1. Emotional information is received by the percipient through mind-to-environment communication.

2. Involves one or more environments and indirect emotional information pertaining to one or more groups of individuals.

3. The source of the information is not a single living organism (e.g. human, animal), but rather the information obtained is in reference to the emotional state

of a group of living organisms of which the emotions of a single individual may be identifiable (e.g. emotions towards community, national health, politics, current events, etc.)

4. Subconscious need for emotional information acquisition present at the time of the experience.

SPECIFIERS

Spontaneous, Intentional; Adaptive, Decisive; Single Episode, Episodic, Continuous; Dream, Intuitive Impression/Emotional (Achievement, Approach, Resignation, Antagonistic, Aesthetic), Compound.

EE3.2 or EE(EI)   **EMPATHIC INTERACTIVE EXPERIENCES**

CRITERIA

1. Emotional information is sent by the agent and received by the subject.

2. Involves one or more individuals and direct emotional information transfer.

3. Subconscious need for emotional information impression present at the time of the experience.

SPECIFIERS

Spontaneous, Intentional; Adaptive, Directive; Suggestive, Compulsive; Single Episode, Episodic, Continuous; Dream, Intuitive Impression/Emotional (Achievement, Approach, Resignation, Antagonistic, Aesthetic), Compound.

EE3.3 or EE(ES)   **EMPATHIC SIMULATIVE EXPERIENCES**

CRITERIA

1. Emotional information is shared between the empath-

ist and one or more participants.

2. Involves one or more individuals and direct emotional information sharing.

3. Subconscious need for emotional information sharing present at the time of the experience.

SPECIFIERS

Spontaneous, Intentional; Adaptive, Directive; Input, Output; Single Episode, Episodic, Continuous; Dream, Intuitive Impression/Emotional, Auditory H, Visual H, Tactile H, Somatic H, Olfactory H, Gustatory H, Compound.

PREFACE

The Classification & Statistical Manual of Extrasensory Experiences (CSM-EE) is a classification of extrasensory experiences with associated criteria designed to facilitate more reliable classification in hopes of one day becoming a standard reference, through future expert review, public commentary, and independent peer review; for clinicians, educators, and researchers challenged with treating, teaching, studying, or investigating into the nature of exceptional experiences. While a comprehensive description of underlying extrasensory processes is not currently possible, it is imperative to highlight that the current classification criteria are detailed descriptions of how extrasensory experiences are expressed and can be recognized by trained professionals.

CSM-EE is intended to assist as a practical, functional, and flexible guide for a wide array of experiences, often viewed as religious or spiritual in nature, that otherwise vary greatly in subjective experience due to varying knowledge and belief systems in experients. CSM-EE assists in identifying like experiences that are similar in phenomenology, but otherwise widely differ in narrative due to culture, language, and religious/spiritual belief. Whereby, the CSM-EE can assist in accurate classification and is therefore a valuable resource for clinicians and students, and a valuable reference for researchers, dealing with extrasensory experiences in a wide diversity of context.

The criteria are concise and fairly explicit and are intended to assist in an objective evaluation of associated phenomenology and experiential presentations in a wide variety of settings by trained professionals. The criteria and associated phenomenological features and specifiers serve in part as a textbook for students who require a well-structured method to understand and classify extrasensory experiences as well as for experienced professionals encountering these experiences for the first time. Criteria and specifiers in this manual were primarily designed to classify the extrasensory experiences of those that are on the initiating end of the experiences (e.g. an agent of telepathic interactive experiences). However, criteria and specifiers can also be utilized to classify the extrasensory experiences of those whom are not on the initiating end of the experiences (e.g. a subject of telepathic interactive experiences).

The criteria and specifiers in this manual were designed to guide clinicians, researchers, and educators in the detailed classification of extrasensory experiences. However, extrasensory experiences can be abstract more often than complex. Because of this, detailed classification will not be an option for all experiences and it is recommended that the user of this manual only use and record specifiers in which are supported by evidence. In other words, if the user is presented with the decision to either vaguely classify an experience or wildly speculate, the user is encouraged to vaguely classify the experience and build upon that classification over time (i.e. as the client/patient reports more experiences). I also hope that the CSM-EE may assist in identifying anomalies in past exceptional experience research and to allow for more informed experimental designs in the future.

Theresa M. Kelly, MSD.

Center for Exceptional Human Experiences

October 24, 2014

INTRODUCTION

The need for a categorical classification system of extrasensory experiences has been clear throughout the history of the field of parapsychology. The *Classification & Statistical Manual of Extrasensory Experiences (CSM-EE)* is a categorical classification designed for clinicians that divides extrasensory experiences into types based on criteria sets with defining features.

Definitions of Extrasensory Perception

Extrasensory Perception (ESP) is defined as: *the reception of information not gained through the recognized physical senses, but sensed by the mind; the acquisition of information about, or response to, an external event, object, or influence (e.g. mental or physical; past, present, or future) otherwise than through any of the known sensory channels.* Extrasensory Perception is assumed the result of the psychical influence of information via an experient's influence over the biological basis of consciousness and the mental process by which we perceive, act, learn, and (i.e. Telepathy) (Kelly, 2011a). In addition, Extrasensory Perception is assumed the result of the psychical influence of our seemingly objective environment/reality, which is presumed as a whole to be a universal information system capable of storing, retaining, and recalling information pertaining to the past and current states of objects and events, and probabilistically determining the potential trajectory of future events (i.e. Clairvoyance) (Kelly, 2011b).

Extrasensory Experiences and Psychopathology

In attempting to compare and differentiate genuine extrasensory experiences to and from psychopathology in the past, clinicians have been uncomfortably confronted with the fact that they were dealing with highly controversial entities that are non-specific; about which there is no general consensus. While a general consensus for psycho-pathology exists, it does lack agreement in regard to which conceptual approach will prove to be more valid in the end (Ullman, 1977). Additional issues include: (1) determining whether an extrasensory experience is a symptom of a mental disorder, (2) if a mental disorder can be caused by extrasensory experiences, and (3) if an individual with a mental disorder is more prone to seeking out, or susceptible to, extrasensory experiences.

While some extrasensory experiences may present similar features to mental disorders, such experiences should not be automatically associated with known psychopathology. Because of this, extrasensory experiences should be classified and addressed in a different manner than mental disorders. Today, the *Diagnostic and Statistical Manual of Mental Disorders* includes a category titled "religious and spiritual problems," V62.89 (Z65.8) which can be utilized:

"[...] When the focus of clinical attention is a religious or spiritual problem. Examples include distressing experiences that involve loss or questioning of faith, problems associated with conversion to a new faith, or questioning of spiritual values that may not necessarily be related to an organized church or religious institution.

According to Lukoff (2000), co-author of the category, the types of religious and spiritual problems covered by this category include the following psychic experiences:

- **Clairvoyance.** Visions of past, future, or remote events.

- **Telepathy.** Communication without apparent physical means.

- **Poltergeist Phenomena.** Physical disturbances in a house with no apparent physical cause.

- **Precognition.** Visions or dreams that provide formerly unknown information.

- **Synchronistic Events.** Meaningful coincidences of two apparently (in terms of cause and effect) non-related events.

While some extrasensory experiences may be viewed by the experient and clinician as not having religious or spiritual features (e.g. when approached from a scientific perspective), this category remains a "catchall" for psychic experiences including psychic healing experiences, out-of-body experiences, auras, medical intuition, communications with spiritual entities, etc. Psychic experiences may be a central feature of an experience or a feature of other types of experiences such as those associated with shamanic practice, kundalini, mystical experiences, and even meditation, which are considered spiritual in nature. To follow, for each extrasensory experience classification, types and subtypes, numeric and alphabetic codes are provided (e.g. 3 or 3.1 – or GE or EC). There are no codes provided for specifiers, as these should be listed in full, or abbreviated, alongside codes.

Example

ICD-9 & 10 & CSM-EE	DSM-5 & CSM-EE
V62.89 (Z65.8) • EE3.1 or EE(EC)	Problems Related to Other Psychosocial, Personal, and Environmental Circumstances (**Religious and Spiritual Problems**) • Extrasensory Experiences (**Empathic Cognitive**) **Specifiers:** moderate, spontaneous, adaptive, episodic, resignation emotions (fear, sadness), people-orientated.

For more information on differentiating extrasensory experiences from mental disorders, please the Center for Exceptional Human Experiences website: www.qpsychics.com/center

Limitations of the Categorical Approach

In the CSM-EE, there is no assumption that each category of extrasensory experience has absolute boundaries dividing it from other extrasensory experiences, as there continues to be existing issues with inter-language, inter-cultural, and inter-individual variants. However, there are standards for basic terminology in the field of parapsychology that have been adopted by the *Parapsychological Association* and the *Parapsychology Foundation*.

The clinician using the CSM-EE should therefore consider that the experiences individuals share are likely to be heterogeneous even in regard to the defining features of the experiences and that boundary cases will be difficult to classify in any way except in a probabilistic manner. This position allows for greater flexibility in the use of the classification system, inspires more precise attention to boundary cases, and stresses the need to acquire additional clinical information that goes beyond classification.

These categories are only applicable when non-extrasensory explanations have been ruled out as the cause of the experience (e.g. cryptomnesia, physical or mental explanations, fraud, psychosis, etc.). Subtypes define equally exclusive and equally exhaustive phenomenological subgrouping within a classification. For example, Telepathy is subtyped based on the expression of mind-to-mind communication and variations in the role of the agent, subject, percipient, and participants, with three subtypes provided: Telepathic Cognition, Telepathic Interaction, and Telepathic Simulation. Specifiers are included in subtypes and provide an opportunity to define a more homogeneous subgrouping of experiences in which share certain features (e.g. Telepathic Cognition, With Spontaneous/Adaptive Intention Features).

Dimensional Approach to Classification

A multidimensional classification system of extrasensory experiences has been proposed. Proposed in this dimensional approach, is that extrasensory experiences can be described utilizing phenomenological dimensions and onset/course dimensions. The former dimensions concern the structural and behavioral nature of the extrasensory experience. The latter dimensions concern the conditions of which bring about an extrasensory experience and influence its course. This multi-dimensional approach was proposed as categorical approaches are limited by at least two key challenges. Firstly, the heterogeneity of extrasensory experiences reflects the possibility that an extrasensory experience may straddle the middle of a continuum between Telepathy, Clairvoyance, and/or Empathy.

Secondly, categorical approaches may fall short in reliability and stability due to briefly specified phenomenological features and perhaps confusion in applying criteria on episodic versus lifetime experiences. This multidimensional, non-prejudice approach to classifying extrasensory experiences allows the clinician to: (1) classify experiences that are subject to associational differences, (2) avoid unjustified labeling and diagnosis, (3) conduct quantitative analyses, and (4) categorize unverified experiences.

For more information on the multidimensional classification system of extrasensory experiences, please visit the Center for Exceptional Human Experiences website: www.qpsychics.com/center

Use of Clinical Judgment

The CSM-EE is a classification of extrasensory experiences that was developed for use in clinical, educational, and research settings. The classification categories, criteria, and textual descriptions are intended to be utilized by individuals with appropriate clinical training and experience, and an appropriate professional education in scientific parapsychology. It is important that the CSM-EE not be applied mechanically by untrained individuals. The specific classification criteria included in CSM-EE are intended to serve as guidelines to be informed by clinical judgment and are not intended to be utilized in a cookbook manner. For example, the exercise of clinical judgment may justify applying a certain classification to an individual's extrasensory experience criteria for classification, as long as the main criteria are present.

On the other hand, lack of familiarity with the field of parapsychology or the contents of the CSM-EE, and/or the excessively adaptive and idiosyncratic application of the CSM-EE criteria substantively reduces its effectiveness as a common language for communication. In addition to the requirement for clinical training and judgment, and a familiarity with the field of parapsychology, the method of data collection is also essential. The valid application of the classification criteria included in this manual necessitates an evaluation that directly accesses the information contained in the criteria sets (e.g. whether an experience is episodic or continuous). Assessments that rely solely on psychological/parapsychological testing not covering the criteria content (e.g. psychical profiling assessments) should not be validly used as the primary source of classification information.

Types of Information in the CSM-EE

The CSM-EE systematically describes each type under the following headings: "Phenomenological Features"; Subtypes and/or Specifiers"; "Associated Research and Laboratory Findings"; "Associated Mental Health Findings"; "Associated Medical Condition Findings"; "Specific Culture, Age, and Gender Features"; "Development and Course"; "Familial Pattern"; and "Differential Classification."

- **Phenomenological Features.** This section clarifies the phenomenological criteria and provides descriptive examples.

- **Subtypes and/or Specifiers.** This section provides definitions concerning applicable subtypes and/or specifiers.

- **Associated Research and Laboratory Findings.** This section provides limited information pertaining to qualitative and quantitative research findings that are associated with a particular type of extrasensory experience.

- **Associated Mental Health Findings.** This section includes mental disorders that are somewhat commonly reported by experients of a particular type of extrasensory experience being discussed, but that are not always present. These disorders may precede, co-occur with, or may be a consequence of the type of extrasensory experience in question.

- **Associated Medical Condition Findings.** This section includes physical medical conditions that are somewhat commonly reported by experients of a particular type of extrasensory experience being discussed, and are not essential to classification. As with associated mental disorders, physical conditions may precede, co-occur with, or may be a consequence of the type of extrasensory experience in question.

- **Specific Culture, Age, and Gender Features.** This section provides guidance for the clinician concerning variations in the presentation of the type of extrasensory experience being discussed that may be attributed to the individual's cultural setting, developmental stage (e.g. childhood, adolescence, adulthood, etc.), or gender.

- **Familial Pattern.** This section describes data on the frequency of the type of experience being discussed among biological relatives of experients of that particular extrasensory experience type in the general population.

- **Associated Terminology.** This section provides the clinician with a wide array of associated terminology used by experients of the type of extrasensory experience being discussed, ranging from parapsychology to popular culture, to assist in classification.

- **Development and Course.** This section describes the typical life-time patterns of presentation and evolution of the type of extra-sensory experience discussed. It contains information on typical *age at onset* and *mode of onset* (e.g. visual or auditory hallucinations) of the type of extrasensory experience discussed.

- **Differential Classification.** This section discusses how to differentiate one type of extrasensory experience from associated exceptional experiences that have similar presenting characteristics.

Associated Experience Prevalence Statistics

On their website, the Institute for Frontier Areas of Psychology and Mental Health (IGPP) states that they receive about 800 requests in the context of "unusual experiences" across Germany every year. They state that ½ of these requests involve needs surpassing a single interview or counseling session. Rather, such needs would be best addressed through continued clinical therapy. Such institutions as the

IGPP, research centers, and educational institutions that are known for their involvement in studying psi and consciousness (e.g. the Koestler Parapsychology Unit at the University of Edinburgh), indicate that the need for clinical attention in the area of unusual experiences is similarly essential. Such demand is not only limited to experients of unusual experiences, but also the experient's relatives seeking advice and social service personal that may be currently assisting the experient or the experient's family in coping with past negative unusual experiences and looking for answers on of to prevent future occurrences.

In a survey conducted by Bauer, Lay & Mischo (1988), taken in various counseling centers in which investigated "occult practices in the teenager population," showed that 79% of the institutions work with these individuals and 75% of the counselors reported feeling inadequately informed to address such cases; with 94% of the requests pertaining to a need for information on the subject of their unusual experience. Therefore, this survey shows that the need for advice and information is not only limited to experients, but also professionals that are confronted with unusual experiences in their practice.

In a data collection study conducted by Coelho, Tierney, & Lamont in 2008, involving the Koestler Parapsychology Unit, University of Edinburgh, found the following:

Age Distribution. 0-9 (4.1%) 10-19 (4.1%) 20-29 (27.6%) 30-39 (10.6%) 40-49 (8.1%) 50-59 (2.4%) 60-69 (0.8%) 70-79 (0.8%). At age group 20-29, 22 out of 34 described their experiences as psychologically, rather than somatically, distressing and these are typically reported as distressing extrasensory experiences.

Gender Distribution. Nearly equal with 44.7% being males, 45.5% being female and 7.3% being multiple gender (i.e. involved more than one person each of different genders e.g. husband and wife). Multiple gender reports typically involved members of the same family.

Purpose of Contact. The majority reason for contacting these institutions was for (1) information 9%, (2) explanation 16%, (3) verification

24%, (4) to describe their experience 6%, and (5) overt requests for help 43%; where 70% were referred to a clinical advisor.

According to Cardeña, Lynn, & Krippner (2000), "Contrary to common belief, hallucinations are not the exclusive province of psychopathology." In a report by Tien (1991), 10-15% of the normal population in the U.S. have had some type of hallucinatory experience within their lifetime. In a report by Verdoux et al. (1998), similar a prevalence was found with 16% in France, and by Poulton et al. (2000), with 13% in New Zealand. Surveys conducted by Sidgwick et al. (1894) and West (1948), that the incidence of hallucinations in the normal population, mainly visual, occur at least once in a single lifespan ranges from 10-14%. According to Schuchter & Zisook (1993), hearing voices is considered a typical sign of the grieving process. According to Grimby (1998), 82% of grieving individuals report some form of communication or "dialogues" with the deceased.

According to Posey & Losch (1983), 71% of a student population reported verbal hallucinations through a questionnaire and according to Barrett & Etheridge (1992), 37% of a student population indicated that they experienced their thoughts aloud. According to Romme and Escher (1989), 70% of the onset of auditory hallucinations occur after a traumatic experience. Lastly, According to an online survey conducted by Landolt et al. (2014), 91% of all participants surveyed reported that they experienced at least 1 exceptional experience. The survey also showed that help-seeking behavior was more frequent in those that had an exceptional experience with a negative valence, and less frequent in individuals without self-reported mental disorders (8.6%) than in individuals with a disorder (35.1%) and preferred to seeing a mental health professional.

Online Enhancements

To keep information as concise as possible and the manual as short as possible for ease of use, a great deal of associated and supporting research was not included in the first edition of the CSM-EE. It was inordinately challenging to determine what information (e.g. statistical, research, phenomenology, etc.) would be covered in the CSM-EE and what would be left out. Because of the scope of research into the extrasensory experiences including both the history of research and the interdisciplinary nature of the subject, and that the CSM-EE will be improved upon and added upon in the years to come to keep up-to-date with future research findings, additional information can be found on the Center for Exceptional Human Experiences website. Professionals are also welcome to submit related research papers, or links to those papers, through the CEHE website for site-inclusion consideration.

If professionals want to contribute to the classification and statistical information in the CSM-EE or suggest improvements to the classification system, please contact the Center for Exceptional Human Experiences through the website. If professionals are looking for more information in regard to extrasensory experiences including differentiating extrasensory experiences from medical symptoms, identifying predisposing, precipitating, perpetuating, and protective factors, implementing measures for purported extrasensory experiences, information of recommended treatment options for negative extrasensory experiences, and current active research indicatives in the area of exceptional human experiences, please visit:

http://qpsychics.com/center

In addition, in early 2015, the University of Alternative Studies will be offering an Online Advanced Professional Certification Program in Scientific Extrasensology, and an Online Master of Science Second-Degree Program in Extrasensory Experiences and Phenomena, both of which will offer training in the professional utilization of the CSM-EE. For more information on these programs, please visit the University's website:

http://qpsychics.com/university/index.html

1. TELEPATHY

Mind-to-Mind Communication

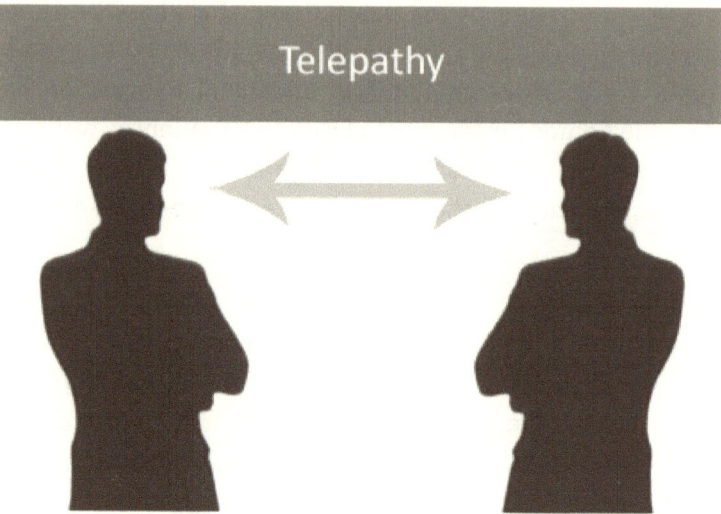

Phenomenological Features

The extrasensory experiences in this section include Generalized Telepathy, Telepathic Cognition, Telepathic Interaction, and Telepathic Simulation. These phenomena have been grouped together to facilitate differential classification of phenomena that include Telepathy as a prominent aspect of the experience. The term *telepathy* was introduced to describe "the communication of impressions of any kind from one mind to another, independently of the recognized channels of sense" (Myers, 1903). Since then, the term has received numerous definitions of which none are utilized exclusively in the scientific community. Popular definitions include the definition of telepathy as "the phenomenologically direct knowledge of another person's thoughts or mental states" (Braude, 1978), and "the paranormal acquisition of information concerning the thoughts, feelings, or activity of another conscious being" (Thalbourne, 2003).

In this manual, telepathy is defined as the psychical influence of thought via experient influence over the biological basis of consciousness and the mental process by which we perceive, act, learn, and re-member; Including mental forms and processes such as the nervous system in which processes and transmits information by electrochemical signaling. Characteristically, people are dynamic information-processing systems whose mental operations can limitedly be described in computational terms as the mind has demonstrated its capacity to store and process visual, auditory, and basic arbitrary packets of information. Experients of telepathic phenomena express influence in regard to the creation, transference, modification, and deletion of single and multiple information packages (Kelly, 2011a).

Subtypes

The following subtypes are phenomenological subgroups exclusive to telepathy only.

1.1 (TC)   **Telepathic Cognition** (see p. 21)
1.2 (TI)   **Telepathic Interaction** (see. p. 28)
1.3 (TS)   **Telepathic Simulation** (see p. 36)

Severity Specifiers

These specifiers should only be used when all criteria for the type or a subtype are currently met. In deciding whether reported experiences should be described as stable/functional, mild, moderate, or severe, the clinician should take into account the number and intensity of the experiences and any resulting impairment in occupational or social functioning.

A. **Stable/Functional.** Intentional experiences of which fit all criteria with few, if any, spontaneous experiences and of which result in no impairment in social or occupational functioning and may or may not increase normal functioning.

B. **Mild.** Few experiences of which fit all criteria and experiences result in no more than minor impairment in social or occupational functioning.

C. **Moderate.** Experiences and functional impairment between "mild" and "severe" are present.

D. **Severe.** Many experiences of which fit all criteria, either episodic or continuous, of which result in marked impairment in social or occupational functioning.

Associated Research and Laboratory Findings

No laboratory findings have been identified that are diagnostic of telepathy. However, a variety of measures from neuroimaging, neuropsychological, and neurophysiological studies have shown differences between groups of individuals with telepathy and appropriately matched control subjects. According to Williams & Roll (2000), in a study examining the correlation between telepathic scoring and alpha abundance, positive correlations have been found. In studies examining the correlation between telepathic scoring and cognitive abilities associated with a particular hemisphere of the brain, mixed results have been found consisting of weak and insignificant evidence. However, EEG studies on two notable psychics suggest right hemisphere processing, but additional brain wave measurement and imaging studies need to be conducted with other notable psychics to make any further determinations.

Numerous studies have implicated the temporal lobe as the region that shapes extrasensory experience. One study has found that individuals with temporal lobe dysfunction reported more extrasensory experiences (psi experiences in general) than other patients. Three studies involving mediums and psychics found elevated temporal lobe signs. Predictions have been made that the hippocampus and amygdala are activated during extrasensory experiences. Further brain regions that may be associated with extrasensory experience are the occipital lobe and the parietal lobe. According to Williams & Roll (2000), predictions have been made that the hippocampus and amygdala are activated during extrasensory experiences because: (1) numerous studies have indicated that extrasensory response consists of implicit emotional memories in which correspond to a perceived object, (2) memory and emotion are processed by these regions.

According to Radin (2006), in two experiments investigating EEG correlations in separated pairs of individuals utilizing a protocol of photic stimulation and EEG measurements, one of which involved two identical twins, followed by 10 replications, 8 of the studies were reported positive. Many replications followed over the years, with one team concluding that the phenomenon could not be easily dismissed and no biophysical mechanism known could account for the correlations. A further replication, where the subject was placed in an fMRI scanner and the agent in a distant room, they found a highly significant increase in brain activity in the subject's visual cortex while the agent was viewing a flickering light. However, while the experiments design is intended to be telepathic, a pre-stimulus or "presponse" (i.e. physiological activity before the stimulus) was found in the subject during experiments suggesting a telepathy/presentiment (i.e. telepathy/clairvoyant) design.

A similar study by Radin (2004), involving the agent viewing a live video image of the subject rather than a flickering light, results overall were somewhat conservative, though 3 out of the 13 pairs of participants independently achieved significant correlations, 5 subjects showed significant EEG peaks, and 10 subjects showed positive EEG peaks. These results "appear to reflect a generalized" subject-agent "relationship." In a series of telephone telepathy experiments by Sheldrake & Smart (2003), (i.e. determining who was calling another after the call was made, but before the caller spoke), resulted in a highly significant effect. A "striking" difference was found when comparing success rates between familiar and unfamiliar callers, where familiar results where significant and unfamiliar were at chance expectation. According to Radin (1997), involving journal articles published between 1966-1973, a total of 450 dream telepathy sessions were reported. These studies range in design, but considering the results of all experiments combined (i.e. meta-analysis), the overall hit rate was above chance expectation.

Specific Culture, Age, and Gender Features

Clinicians assessing beliefs and claims in socioeconomic or cultural situations that are dissimilar from their own must take cultural dissimilarities into account. Ideas that may appear to be questionable or even delusional in one culture or subculture (e.g. Buddhists, New Agers, Spiritualists, Wiccan Practitioners, and those who engage in regular meditative practices) may be commonly believed in another. In some cultures, telepathic hallucinations with spiritual or religious content may be a normal part of spiritual or religious experience (e.g. abhijna; Buddhist "mind-penetrating" knowledge, mothers intuition; suggestive of an empathic or telepathic connection between mother and child often seen in a spiritual context). These varying beliefs may have subtle to blatant differences in terminology and descriptions leaving the clinician with the difficult task of properly categorizing experiences into parapsychological types and subtypes.

In regard to physical location, in a study conducted by Haraldsson & Houtkooper (1991), individuals in the U.S. reported telepathic experiences 54%, which far exceeded any of the European nations surveyed, including Italy with 41%, Finland with 40%, West Germany with 39%, Great Brittan with 36%, France and Iceland with 34%, Holland with 29%, Sweden with 24%, Spain and Belgium with 21%, Ireland with 19%, Norway with 17%, and Denmark with 15%. The total percentage for all European countries was 34%. Only one Asian country was included in this study, which was South Korea with 48%. This study included reports for clairvoyance and contact with the dead, where telepathy was reported more frequently than the other two types of experiences. The study suggests that "weighing the figures by national population sizes, it can be estimated that 32% of Europeans from these countries have experienced telepathy" along with 54% of Americans.

Initial experiences (onset) of telepathic phenomena typically occur within the first several years after birth and/or during puberty. Early onset may involve several spontaneous experiences of which may or may not affect the child psychologically, emotionally, or socially. Experiences in which have an early onset and continue throughout life without extended pause (e.g. 1 year or more without an experience) typically remain stable/functional in the long term. In some generalized ESP experiments, children tend to score higher than adolescents and adults. However, many similar studies have been unsuccessful in in demonstrating age dependent differences in scoring (Palmer, 1978).

The onset of telepathic phenomena during puberty, most common between the ages of 13-16, is typically induced to compensate for an inability to effectively communicate their wants, needs, and/or thoughts verbally. Experients may feel they have had a recent decline in quality of life, academic performance, and/or social relationships. During this time experiences are typically spontaneous, and can range from mild to severe depending on the severity of needs the experient feels they are unable to communicate and obtain. Experiences in which are moderate to severe have a high probability of continuing in severity unless the want or needs of the adolescent are met. Onset during this age rage may also be induced by another individual unwilling to meet the experients wants or needs. In other words, the experient is communicating effectively, but the recipient does not understand effectively (e.g. a parent that does not understand a child's limitations due to developmental or behavioral issues), or simply refuses to meet the experients needs (e.g. a bully at school, or an abusive or neglectful parental figure).

Adult onset may occur at any age and is typically precipitated by and inability to verbally communicate wants, needs, or thoughts, or a desire to deepen a mental and emotional connection with another individual(s) (e.g. spouse, children). Adult onset is typically stable/functional to mild unless precipitated by experiences that amount to trauma, illness, or any other type of sudden uncomplimentary experience, acute or chronic, that results in a major disturbance in the experient's life. In the case of the latter, moderate to severe experiences are typically common. Spontaneous experiences are common regardless of the severity. However, stable/functional to mild experiences are more likely to be the product of intention, while moderate to severe experiences are mainly spontaneous.

Gender differences have been the focus of some studies. Overall, there appears to be no clear trend for differential scoring between males and females (Palmer, 1978). However, in an online survey conducted by Parra (2001), that related to gender and age found that women tended to report relatively higher numbers of telepathic experiences compared to men, that men tended to show more negative emotional impact compared to women, and that for birth order, it was found that those who were 'only' children tend to experience a greater number of paranormal experiences in general than compared with those who have siblings. In Europe, 38% of women surveyed reported telepathic experiences with men surveyed reporting only 30%, and there was a comparable result in the U.S. where 59% of women surveyed reported telepathic experiences with men surveyed reporting only 47% (Haraldsson & Houtkooper, 1991). In addition, there has been evidence supporting that mixed-gender pairings (agent and subject) are more successful than same-gender pairings (Dalton & Utts, 1995).

Familial Patterns

Occasionally one biological parent or grandparent of an experient of telepathic phenomena reports a history of telepathic-like experiences. Familial patterns most common are telepathic experiences between mother and child, spouses/lovers, identical twins, and occasionally between fraternal twins, siblings, and meditation partners. In regard to marital status in Europe, telepathic experience reports are 31% married couples, 36% individuals (single), 44% living as married (but not legally married), 53% separated, and 47% divorced. A similar trend was found in the U.S. where telepathic experiences are reported 51% by married couples, 53% by individuals (single), 65% by those living as married (but not legally married), 66% separated, and 64% divorced.

Overall, comparatively fewer single and married individuals report telepathic experiences than the "combined broken-relationship group" (i.e. living as married, separated, or divorced) (Haraldsson & Houtkooper, 1991).

Associated Terminology

Emotional content. Experients of telepathy in which primarily sense emotional content, but still receive more than emotional content on occasion, may use the following terminology: empath • empathic • empathist • empathy • gut feeling • intuition • intuitionism • intuitive • intuitvism • keen intuition • mothers intuition • sensitive • tele-empathic • tele-empathy.

Visual content. Experients of telepathy in which primarily sense visual content may use the following terminology: psipath • psipathic • telepathic • telepathist • telepathy.

Additional terminology. Used in a context involving the experient and at least one other individual, including both telepathy and telepathy-like terminology: anomalous communication • audible thoughts • insight • mental compulsion • mental suggestion • mind control • mind influence • mind/mental rape • mind reader/reading • psychic communication • psychic knowledge • second sight • six sense • thought broadcasting • thought insertion • thought reception • thought transference • thought transmission • thought withdrawal • twin telepathy • second sight • six sense.

Criteria for Telepathic Experiences

A. **Characteristic phenomenology:** all of the following are required criteria for classification.

    (1) Mind-to-mind communication.

    (2) Involves two or more individuals.

(3) The agent and subject, percipient, or all participants, are living organisms.

B. **Social/occupational need:** A subconscious need has been identified as the catalyst for the initiation of telepathic processes (i.e. identified an inability to communicate wants, needs, or thoughts to an individual(s) in an interpersonal, academic, or occupational context).

C. **Validation:** The experience has been validated by an individual other than the experient (e.g. the subject(s) confirmed the accuracy of the information received by the experient), and/or the clinician determines the experience was more than a coincidence/chance occurrence based on the quality of the information received and reported, and all other possible explanations for obtaining the information is excluded. If validation does not apply, yet telepathic processes are still plausible, the experience should be classified as "Possible Telepathy" (PT).

D. **Empathy Exclusion:** Psychical Empathy has been ruled out because more than emotional content was involved in the experience(s).

E. **Clairvoyance Exclusion:** Clairvoyance has been ruled out because mind-to-mind communication has been identified as the basis of the experience(s).

Telepathy Subtypes

The subtypes of Telepathy are defined by the predominant phenome-nology of reports. The determination of a particular subtype is based on the clinical picture that occasioned the most recent experiences, and may therefore change over time. Not infrequently, the description of experiences may include phenomena that are characteristic of more than one subtype. The choice among subtypes depends on the follow-ing algorithm: Telepathic Cognition (TC) is assigned whenever infor-mation is telepathically acquired by the telepathist originating from a subject; Telepathic Interaction (TI) is assigned whenever information is telepathically acquired by a subject originating from the telepathist; Telepathic Simulation (TS) is assigned whenever information is shared between the telepathist and a participant; If two or three subtypes are assigned, all should be listed; Generalized Telepathy (GT) is assigned when all subtypes appear to apply (optional), or a clear choice is una-ble to be made, but appears to suggest only telepathic phenomena. In addition, when a clear choice cannot be made, the clinician should consider a dimensional approach to classifying the experiences.

1.1   Telepathic Cognition (TC)

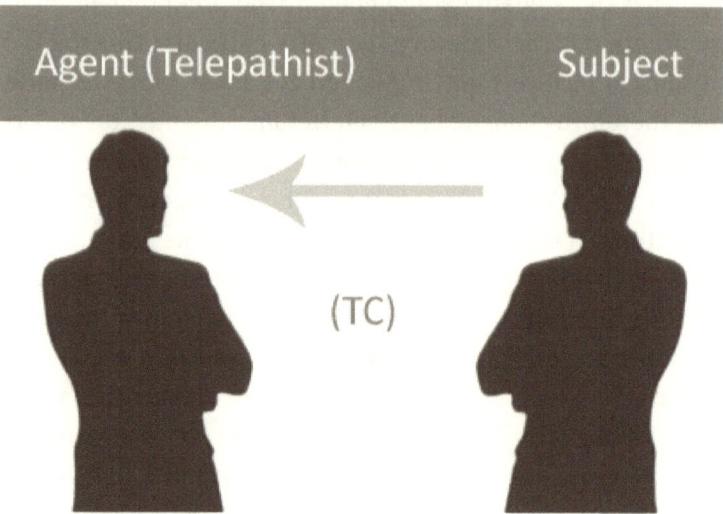

Phenomenological Features

The essential feature of the (TC) type of telepathy is *the phenomeno-logically direct knowledge of another person's thoughts or mental states* (Braude, 1978). In cases of Telepathic Cognition, one individual is retrieving information from another, i.e. one individual is able to "pick up on" the thoughts of another individual. The person from whom the thoughts originate does not play and intentional part in the information teleportation processes. Instead, the process is assumed entirely evoked by the receiver/percipient. In other words, in regard to telepathic cognition, the "receiver" is the telepathist (i.e. an individual capable of evoking telepathic processes). Here, the telepathist will become aware of the other individual's mental state or states, but should be able to clearly identify that the thought did not originate in their own mind. Here the information is received and perceived by the telepathist, but the thought did not develop from a chain of prior thoughts belonging to the telepathist. Instead, the thought appears to "pop up," but is immediately associated with a specific individual other than the telepathist, or simply identified as not originating from the telepathist.

Hallucinations that may occur include any sensory modality (e.g. visual, auditory, olfactory, gustatory, and tactile), but visual and auditory hallucinations are the most common. Hallucinations occur while the experient is awake or in an altered state of consciousness (e.g. hypnagogic, hypnopompic, or a trance state). Hallucinations are typically unobtrusive unless exacerbated by stress or a mental disorder (e.g. an anxiety disorder may result in the disorganization of subconscious needs and goals resulting in the over stimulation of telepathic processes) (Braude, 1978; Kelly, 2011a).

Intention Specifiers

The first set of specifiers is for identifying whether the experience was intentional or unintentional.

A. **Spontaneous.** This specifier applies when information appears to "pop into mind" rather than being intentionally requested by the percipient.

B. **Intentional.** This specifier applies when a percipient selects or specifies another individual from whom they wish to extract information. However, this specifier also applies when a percipient generalizes their search from "by whom," to "by what" type of information the percipient needs. In the case of the latter, one or several subjects may be perceived.

The second set of specifiers is for identifying the subconscious or conscious need or goal that is assumed to be the catalyst for initiating telepathic cognitive processes.

A. **Adaptive.** This specifier applies when information acquisition is initiated to assist the percipient in understanding and adapting to subjects in which they typically have some level of emotional investment.

B. **Decisive.** This specifier applies when information acquisition is initiated to assist the percipient in coming to a decision involving subjects in which they typically have some level of emotional investment.

Development and Course

Childhood onset may present itself through dreams, visual and/or auditory hallucinations, with intuitive impressions (i.e. gut feelings, intuition) being also common. Adolescent onset primarily presents itself through visual and/or auditory hallucinations with telepathic dreams and intuitive impressions (i.e. gut feelings, intuition) being also common. However, other types of hallucinations (e.g. olfactory, tactile, etc.) are less common. Adult onset primarily presents itself through telepathic dreams, intuitive impressions, or during crisis situations in the form of hallucinations subconsciously deem most appropriate for notification. Compound modalities are more common amongst identical and fraternal twins.

Course Specifiers

These specifiers are for identifying the characteristic course of telepathic cognitive experiences over time.

A. **Single Episode.** This specifier applies when the percipient experiences a telepathic dream, or impression, or hallucination without a prior history of episodes.

B. **Episodic.** This specifier applies when the percipient experiences telepathic dreams, or impressions, or hallucinations of which seem to occur irregularly and of which the duration of the experience is very momentary. An episodic hallucination may involve a quick flash of an image or an auditable single word or short phrase with the duration of the experiencing lasting only a maximum of a couple of seconds. An episodic hallucination may also involve a more "movie-like" or dynamic image or auditable whole sentences or rhymes (e.g. songs) with the duration typically lasting no longer than a few seconds. While the percipient may appear distracted during a telepathic episode, the experient should still be fully aware of their surroundings.

C. **Continuous.** This specifier applies when the percipient experiences telepathic impressions or hallucinations of which seem to occur in a continual manner, or when episodes are so frequent it is difficult for the percipient to determine where one episode ends and another begins (e.g. prolonged and closely spaced episodes).

Modality Specifiers

These specifiers are for identifying the characteristic mode(s) of a telepathic experience. In any case, some emotional investment in the individual, or the situation in which the individual resides, on the experients behalf is expected.

A. **Dream.** Refers to telepathic information acquisition during sleep.

B. **Intuitive Impressions/Emotional.** Refers to non-hallucinatory sensations of which can be described as telepathically received emotional content.

C. **Auditory Hallucinations.** Hallucinations of hearing/sound. Typically only involves verbal hallucinations as opposed to non-verbal hallucinations. While the origin of telepathic auditory hallucinations are external, they are typically perceived as internal (i.e. heard within the mind as opposed to seemingly heard by the physical ear), and stem from an identifiable location (i.e. a subject in spatial proximity to the percipient, or a subject at a distance).

D. **Visual Hallucinations.** Hallucinations of sight. Involving a perceived complexity classified as simple or complex. If the entire environment is replaced by the visual hallucination, the hallucination is classified as a scenic or panoramic hallucination. Visual hallucinations in which are located beyond the visual field (e.g. in the back of the mind, third eye vision, etc.) are classified as extra-campine hallucinations. Using the perceived shape of the hallucination, visual hallucinations can be classified as formed, organized, or unformed (i.e. abstract).

E. **Tactile Hallucinations.** Hallucinations of pressure and touch. Can include a wide range of sensations from a pat on the shoulder, a knee injury, a blow to the head, and hot and cold sensations. Tactile hallucinations are classified based on the type of sensation experience (e.g. painful sensations are classified as pain hallucinations; temperature sensations are classified as thermal/thermic hallucinations).

F. **Somatic Hallucinations.** Hallucinations from inside the body (e.g. heart, lungs, sensations within the limbs, stomach e.g. nausea). Also known as somatosensory hallucinations.

G. **Olfactory Hallucinations.** Hallucinations of smell. These hallucinations are typically extrinsic where the localization of the smell is outside of the body (e.g. the smell of tobacco, fumes from a fire, flowers and grass in a park, the perfume of a loved one, etc.).

H. **Gustatory Hallucinations.** Hallucinations of taste. May include a wide range of taste sensations classified as bitter, sour, sweet, 'disgusting,' etc., but can be classified in more specific terms (e.g. tobacco, garlic, salt, blood, etc.).

I. **Compound.** Several modalities are involved, in which case each mode involved should be noted.

Associated Mental Health Findings

Mental health disorders somewhat common in experients of Telepathic Cognition include: Alcohol and/or Substance Abuse/Dependence; Attention Deficit/ Hyperactivity Disorder; Depressive Disorder; Generalized Anxiety Disorder; Panic Disorder with or without Agoraphobia; and Social Phobia (Kelly, 2011a).

Associated Medical Condition Findings

Physical medical conditions somewhat common in experients of Telepathic Cognition can include: Asthma; Allergies; Migraines; and occasionally a history of Cancer (e.g. lung, breast, chest area in general.) (Kelly, 2011a).

Differential Classification

A wide variety of extrasensory phenomena can present with similar phenomenology. These include:

o **Empathy**. Applied when there is evidence to support that emotional content is the only type of content perceived by the percipient. However, if other informational content is involved, the experience should be classified as telepathy.

o **Telepathic Interaction**. Applied when there is evidence to support that the telepathist is only capable of one-way, telepathist-to-subject communication in the form of dual independent thought and impression.

o **Telepathic Simulation**. Applied when there is evidence to support that the telepathist is capable of two-way telepathist-to-participant and participant-to-telepathist communication in the form of shared and identical feelings, thoughts, or behaviors.

o **Clairvoyance**. Applied when there is evidence to support that information is obtained 'about' a subject, but the information obtained is not 'from' the subject. Information received telepathically is typically in first-person, second-person, or 'direct,' while information received clairvoyantly is typically in third-person or 'indirect' (e.g. if the percipient receives information described "as though they are looking through the eyes of another individual," this would be classified as telepathy. However, if the percipient describes receiving the information "as though they are looking at the individual and the individual's surroundings," this would be classified as clairvoyance. In a similar circumstance, if an individual becomes aware of an ailment in their own body, or the body of another individual, but no other individual was aware of the physical ailment, then this would be classified as clairvoyance. This is because (1) telepathy is mind-to-mind communication, not mind-to-body communication, (2) telepathy must include at least two individuals, and (3) because the knowledge of the ailment did not originate from another mind.)

o   **Mediumship**. Applied when there is evidence to support that information is obtained from a non-physically living being; as telepathy only refers to the communication of two living organisms. That is, living in the sense of existing within a physical, corporeal body (e.g. if the percipient receives information described as originating from a discarnate entity, i.e. the spirit of a deceased individual, or another form of entity that is not a physically living being such as a "Spirit Guide" or "Angel," this would be classified as mediumship in general, clairvoyant interaction, or clairvoyant cognition.)

Criteria for Telepathic Cognitive Experiences

A. **Characteristic phenomenology:** all of the following are required criteria for telepathic cognitive experiences including criteria for telepathy in general.

(1) Information is received by the percipient.

(2) Information received is in first person perspective (e.g. If visual: the image received is from the subject's perspective), or narrative (e.g. If auditory: the words received are from the subject's perspective "He/She is happy today;" or occasionally "'I am happy today;" with additional information in reference to the subject).

(3) Subconscious need for information acquisition present at the time of the experience.

1.2 Telepathic Interaction (TI)

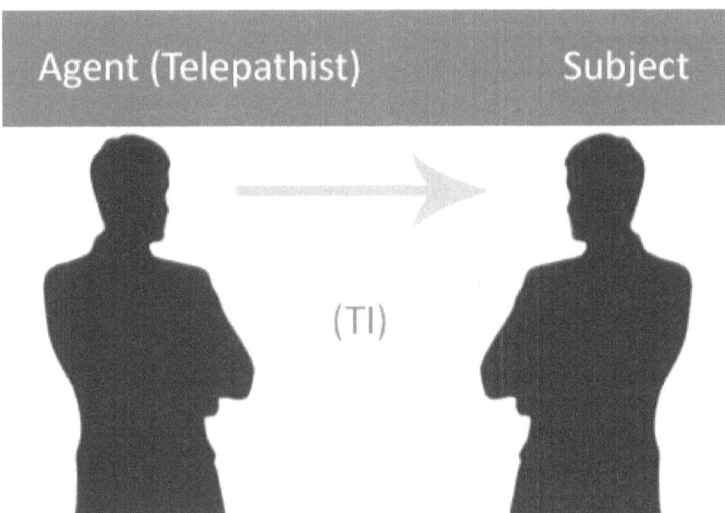

Agent (Telepathist)   Subject

(TI)

Phenomenological Features

The essential feature of the (TI) type of telepathy is *the causal influence of one mind on another without the intervention of the five senses* (Braude, 1978). In cases of telepathic interaction, individuals who engage in telepathic interaction (agents) appear to do so via commands based on the agent's subconscious need to have the individual (subject) feel or behave in a particular way. However, it appears that "hypnogenic telepathic Interaction" is involved in a manner producing a mild hypnotic state in the subject via an agent's command to do so. This is void of the agent having to mentally produce any feelings associated with a hypnotic state within him/herself (the telepathist). This method of telepathic interaction is also classified as *"hypnotic telepathic interaction"* or *"hypnotic telepathy"* that appears to not only evoke strong emotions in the subject, but also typically results in an action on the subject's behalf more often than the method associated with simply "commanding" an act.

Therefore, hypnotic telepathic interaction appears to be the strongest form of telepathy and the most dangerous, raising an assortment of moral and ethical questions as to how such an ability should be utilized in practical applications. Further studies suggest that initial telepathic "impressions" (i.e. commands or evoked feelings) do not always fade away with time. Rather, some initial impressions occasionally result in the same strength of emotions or "need to act" anytime (1) associated images of the agent, or (2) associated feelings pertaining to the feelings initially evoked by the agent, are mentally accessed (e.g. feelings of obsession, i.e. often misconstrued as love, may be triggered by the subject visually thinking about the agent, seeing the agent in person, or simply thinking about other loved ones) (Braude, 1978; Kelly, 2011a).

Intention Specifiers

The first set of specifiers is for identifying whether the experience was intentional or unintentional.

A. **Spontaneous.** This specifier applies when the agent impresses the subject void of the agent's conscious choice, decision, or intention.

B. **Intentional.** This specifier applies when an agent selects or specifies a subject to be impressed with information or coerced into action or into the expression of a type of behavior. However, this specifier also applies when an agent generalizes their interaction from "who" to "what" the agent needs from the subject to result in a modification of a situation involving the subject. In the case of "what," one or more subjects may be impressed.

The second set of specifiers is for identifying the subconscious or conscious need or goal that is assumed to be the catalyst for initiating telepathic interactive processes.

A. **Adaptive.** This specifier applies when information impression is initiated to assist the subject in understanding and adapting to the agent's needs or goals. Here the agent typically has some level of emotional investment in the subject or the situation in which the subject occupies.

B. **Directive.** This specifier applies when information impression is initiated to assist the subject in complying with a suggestion or command, from the agent, to act or behave in a specific or generalized manner. Behaviors can range from <u>common</u> to <u>unusual</u> and <u>acceptable</u> to <u>unacceptable</u> in regard to social norms. Social actions include <u>rational</u> (i.e. the action leads to a valued goal, but with no thought of its consequences and often without consideration of the appropriateness of the means), <u>instrumental</u> (i.e. actions which are planned and carried out after evaluating the goal in relation to other goals, and after thorough consideration of various means and consequences to achieve said goals), and <u>affectional</u> (i.e. actions which are carried out due to one's emotions, to express personal feelings). Here the agent typically has some level of emotional investment in the subject or the situation in which the subject occupies.

The third set of specifiers is for identifying the type of impression experienced by the subject.

A. **Suggestive.** This specifier applies when the information impressed can be identified as a proposal of which may be accepted or rejected per the subject's prerogative.

B. **Compulsive.** This specifier applies when the information impressed can be identified as coercive towards either irrational behavior, or coercive towards rational behaviors, but the subject feels they are behaving against their will or by force.

Development and Course

Childhood onset may present itself through a subject's dreams, visual and/or auditory hallucinations, with intuitive impressions (i.e. gut feelings, intuition, emotional content) also being common. Adolescent onset primarily presents itself through the subject experiencing visual and/or auditory hallucinations, with telepathic dreams, and intuitive impressions (i.e. gut feelings, intuition, emotional content) also being common. However, other types of hallucinations (e.g. olfactory, tactile, etc.) are less common, with hypnotic telepathy being more common. Adult onset primarily presents itself through telepathic dreams, intuitive impressions, or during crisis situations in the form of hallucinations or strong emotional content subconsciously deem most appropriate for suggestion/compulsion. Compound modalities are more common amongst identical and fraternal twins.

Course Specifiers

These specifiers are for identifying the characteristic course of telepathic interactive experiences over time.

A. **Single Episode.** This specifier applies when the agent impresses a telepathic intuitive impression or a hallucination onto the subject(s) and the agent and subject report no prior history of episodes. This specifier also applies when the agent suggests or compels a subject(s) to engage in an action or behavior that the subject reports is not typical (i.e. the subject has not responded in such a way in similar circumstances in the past), but the agent and subject report no prior history of episodes.

B. **Episodic.** This specifier applies when the agent impresses telepathic emotional content, intuitive impressions, or hallucinations onto a subject(s) of which seem to occur irregularly and of which the duration of the experience is very momentary. An episodic hallucination may involve a quick flash of an image or an auditable single word or short phrase with the duration of the experience lasting only a maximum of a couple of seconds. An episodic hallucination may also involve a more "movie-like" or dynamic image or auditable whole sentences or rhymes (e.g. songs) with the duration typically lasting no longer than a few seconds. This specifier also applies when the agent suggests or compels the subject(s) to engage in an action or behavior that is not typical of the subject(s) in the past, but now the subject(s) irregularly acts or behaves in such a manner.

C. **Continuous.** This specifier applies when the agent impresses tele-pathic emotional content, intuitive impressions, or hallucinations onto a subject(s) of which seem to occur in a continual manner, or when episodes are so frequent it is difficult for the experient or subject to determine where one episode ends and another begins (e.g. prolonged and closely spaced episodes).

Modality Specifiers

These specifiers are for identifying the characteristic mode(s) of a tel-epathic interactive experience. In any case, some emotional invest-ment in the subject, or the situation in which the subject resides, on the experients behalf is expected.

A. **Dream.** Refers to telepathic impression during sleep where the subject is impressed with information during the dream state and/or engages in an action or behavior once awake due to the impressed content of the dream.

B. **Intuitive Impressions/Emotional.** Refers to non-hallucinatory sen-sations of which can be described as emotional content telepathi-cally impressed onto a subject(s) that may result in an action or behavior.

C. **Auditory Hallucinations.** Hallucinations of hearing/sound. Typical-ly only involves <u>verbal</u> hallucinations as opposed to non-verbal hal-lucinations. While the origin of telepathic auditory hallucinations are external, they are typically perceived as <u>internal</u> (i.e. heard within the mind as opposed to seemingly heard by the physical ear), and stem from an identifiable location (i.e. a subject in spatial proximity to the agent, or a subject at a distance).

D. **Visual Hallucinations.** Hallucinations of sight. Involving a per-ceived complexity classified as <u>simple</u> or <u>complex</u>. If the entire en-vironment is replaced by the visual hallucination, the hallucination is classified as a <u>scenic</u> or <u>panoramic</u> hallucination. Visual halluci-nations in which are located beyond the visual field (e.g. in the back of the mind, third eye vision, etc.) are classified as extra-campine hallucinations. Using the perceived shape of the hallucination, visual hallucinations can be classified as formed, organized, or unformed (i.e. abstract).

E. **Tactile Hallucinations.** Hallucinations of pressure and touch. Can include a wide range of sensations from a pat on the shoulder, a knee injury, a blow to the head, and hot and cold sensations. Tactile hallucinations are classified based on the type of sensation experience (e.g. painful sensations are classified as pain hallucinations; temperature sensations are classified as thermal/thermic hallucinations).

F. **Somatic Hallucinations.** Hallucinations from inside the body (e.g. heart, lungs, sensations within the limbs, stomach e.g. nausea). Also known as somatosensory hallucinations.

G. **Olfactory Hallucinations.** Hallucinations of smell. These hallucinations are typically extrinsic where the localization of the smell is outside of the body (e.g. the smell of tobacco, fumes from a fire, flowers and grass in a park, the perfume of a loved one, etc.).

H. **Gustatory Hallucinations.** Hallucinations of taste. May include a wide range of taste sensations classified as bitter, sour, sweet, 'disgusting,' etc., but can be classified in more specific terms (e.g. tobacco, garlic, salt, blood, etc.).

I. **Compound.** Several modalities are involved, in which case each mode involved should be noted.

Associated Mental Health Findings

Mental health disorders somewhat common in experients of telepathic interaction include: Alcohol and/or Substance Abuse/Dependence; Attention Deficit/ Hyperactivity Disorder; Depressive Disorder; Disruptive Behavior Disorder; Obsessive Compulsive Disorder; Sleep Disorder (Kelly, 2011a).

Associated Medical Condition Findings

Physical medical conditions somewhat common in experients of tele-pathic interaction can include: Hypertension or Hypotension; Chronic Fatigue Syndrome; Chronic Pain (e.g. Myalgia, Fibromyalgia); Blood Disorders (e.g. Anemia); Digestive Disorders; Palpitations (Kelly, 2011a).

Differential Classification

A wide variety of extrasensory phenomena can present with similar phenomenology. These include:

o **Empathy**. Applied when there is evidence to support that emo-tional content is the only type of content perceived by the subject. However, if other informational content is involved, the experi-ence should be classified as telepathy.

o **Telepathic Cognition**. Applied when there is evidence to support that the percipient is the agent and that the telepathist is only ca-pable of one-way, subject-to-telepathist communication in the form of dual independent thought and the acquisition of knowledge.

o **Telepathic Simulation**. Applied when there is evidence to support that the subject is a participant and that the telepathist is capable of two-way telepathist-to-participant and participant-to-telepathist communication in the form of shared and identical feelings, thoughts, or behaviors.

o **Clairvoyance**. Applied when there is evidence to support that the information obtained by the subject 'about' an alleged agent was not 'from' the alleged agent. Information received telepathically is typically in first-person, second-person, or 'direct,' while infor-mation received clairvoyantly is typically in third-person or 'indi-rect' (e.g. if the subject receives information described "as though they are looking through the eyes of another individual," this would be classified as telepathy. However, if the alleged subject describes receiving the information "as though they are looking at the alleged agent and the alleged agent's surroundings," this experience would be classified as clairvoyance with the percipient classified as the agent rather than the subject.

o **Mediumship.** Applied when there is evidence to support that information was obtained by an alleged subject from a non-physically living being; as telepathy only refers to the communication of two living organisms. That is, living in the sense of existing within a physical, corporeal body (e.g. if the percipient receives information or is physically directed, and the origin of the information or direction is from a discarnate entity, i.e. the spirit of a deceased individual, or another form of entity that is not a physically living being such as a "Spirit Guide" or "Angel," this would be classified as mediumship in general, clairvoyant interaction, or clairvoyant cognition.)

Criteria for Telepathic Interactive Experiences

A. **Characteristic phenomenology:** all of the following are required criteria for telepathic interactive experiences including criteria for telepathy in general.

(1) Information is sent by the agent.

(2) Information is sent in second-person perspective (e.g. If visual: the image impressed is from the agent's perspective) or narrative (e.g. If auditory: the words impressed are from the agent's perspective i.e. "You want to behave this way.").

(3) Subconscious need for information impression present at the time of the experience.

1.3   Telepathic Simulation (TS)

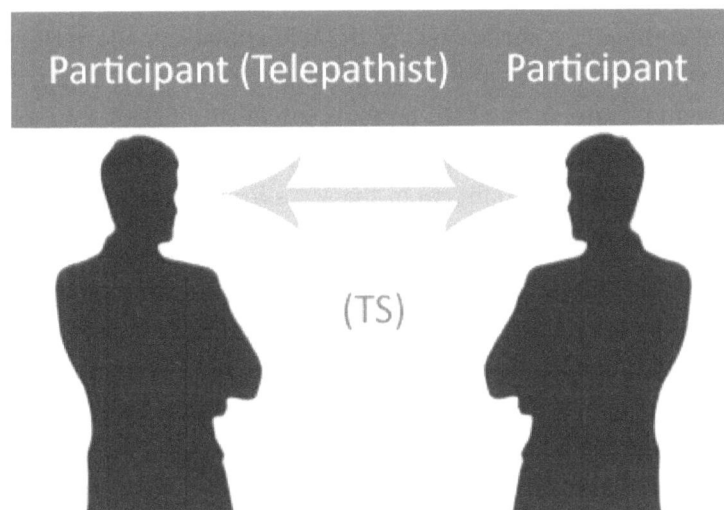

Phenomenological Features

The essential feature of the (TS) type of telepathy is *a case in which an individual's mental state appears to produce a similar mental state in someone else* (Braude, 1978). In other words, the telepathist's mental state produces a similar mental state in a participant. Through this type of telepathy, the telepathist does not "know" telepathically what the mental state of a participant is, nor is the information impressed, but rather it appears that the mental states of the telepathist and participant instantaneously become qualitatively identical. The identical-ness of the mental state is debatable, as there is no empirical evidence to support this at this time. However, reports in regard to this type of telepathy suggest exact, or nearly exact, mental states rather than more associative states (e.g. a star for a star, rather than a star for a pinwheel or daisy).

This type of telepathy also appears to be more non-invasive as participants are typically unaware that, or do not "know" that, the mental state is "not their own," as it appears to be less intrusive than other types of telepathy. The skilled telepathist would however be able to identify that the simulated mental state originated from him/herself if the telepathist intentionally shared information with a participant. In other words, the telepathist can share his/her own mental state with a participant, or the telepathist can evoke the sharing process of a participant's mental state to replace the telepathist's own mental state. In the end, perhaps the most efficient way to view telepathic simulation is as though the mental states have been shared through the exact transmission of the state from the telepathist to the participant or from the participant to the telepathist.

Another feature that appears exclusive to telepathic simulation is that of networking. In this regard, the telepathist is not only able to interact with one participant, in which is typical of telepathic impressionists and semi-typical of telepathic cognitives, but rather 'network' their ideas to many participants at the same time. In other words, telepathic simulators typically work with *group participant mental state sharing* just as effectively as *single participant state sharing* (i.e. the telepathist can get everyone "on the same page" at once) (Braude, 1978; Kelly, 2011a).

Intention Specifiers

The first set of specifiers is for identifying whether the experience was intentional or unintentional.

A. **Spontaneous.** This specifier applies when the telepathist initiates the telepathic sharing of information void of conscious intent.

B. **Intentional.** This specifier applies when the telepathist intentionally specifies who is to participate in the sharing process, or what information will be shared. If the process involves 'what' rather than 'who,' participants may be selected subconsciously by the telepathist based on their relativity to the required result.

The second set of specifiers is for identifying the subconscious or conscious need or goal that is assumed to be the catalyst for initiating telepathic simulative processes.

A. **Adaptive.** This specifier applies when information shared is initiated by the telepathist to assist participants in understanding and adapting to the telepathist's, or group's and the telepathist's needs or goals. The most common goal is to provide <u>social or emotional comfort</u> and/or a <u>sense of security</u>. Here the telepathist and participants typically have some level of emotional investment in each other or the situation in which they occupy.

B. **Directive.** This specifier applies when information shared is initiated by the telepathist to assist participants in an action towards a goal (i.e. motivation). In other words, to provide purpose and direction to behavior. Behaviors can range from <u>common</u> to <u>unusual</u> and <u>acceptable</u> to <u>unacceptable</u> in regard to social norms. Social actions include <u>rational</u> (i.e. the action leads to a valued goal, but with no thought of its consequences and often i.e. without consideration of the appropriateness of the means), <u>instrumental</u> (i.e. actions which are planned and carried out after evaluating the goal in relation to other goals, and after thorough consideration of various means and consequences to achieve said goals), and <u>affectional</u> (i.e. actions which are carried out due to one's emotions, to express personal feelings). However, while emotion does appear to be involved in telepathic simulation, it does not appear to penetrate the barrier of self-control like telepathic interaction. Here the telepathist and participants typically have some level of emotional investment in each other or the situation in which they occupy.

The third set of specifiers is for identifying the direction of the telepathic simulative experience.

A. **Input**. This specifier applies when the participant shares information with the telepathist. Here information in regard to the participant has been shared with the telepathist (e.g. the telepathist was feeling anxious, but the participant was not feeling anxious prior to simulation; however, post simulation, neither participant felt anxious).

B. **Output**. This specifier applies when the telepathist shares information with the participant. Here information in regard to the telepathist has been shared with the participant (e.g. the participant was feeling anxious, but the telepathist was not feeling anxious prior to simulation; however, post simulation, neither participant felt anxious).

Development and Course

Childhood onset may present itself through simulated dreams, simulative intuitive impressions (i.e. gut feelings, intuition, emotional content), and simulated visual and/or auditory hallucinations being also common. Adolescent onset primarily presents itself through simulated dreams, simulative intuitive impressions (i.e. gut feelings, intuition, emotional content), and simulated visual and/or auditory hallucinations being also common. However, other types of simulative hallucinations (e.g. olfactory, gustatory, etc.) are less common with tactile and/or somatic simulative hallucinations being more common. Adult onset primarily presents itself through simulative dreams, simulated intuitive impressions, or during crisis situations in the form of hallucinations or strong emotional content subconsciously deem most appropriate in assisting adaptive or directive behavior (i.e. motivation). Compound modalities are more common amongst identical and fraternal twins.

Course Specifiers

These specifiers are for identifying the characteristic course of telepathic simulative experiences over time.

A. **Single Episode.** This specifier applies when the telepathist initiates the sharing process of a telepathic intuitive impression or a hallucination with a participant and the telepathist reports no prior history of episodes. This specifier also applies when a the telepathist (i.e. output) or participant (i.e. input) shares information to provide purpose and direction to behavior in which the telepathist or participant report as not typical (i.e. a participant has not responded in such a way in similar circumstances in the past). The classification of a single telepathic simulative experience can be difficult as it is often challenging to identify who is playing the role of the telepathist (i.e. if the individual reporting the experience is the initiator of telepathic simulative processes).

B. **Episodic.** This specifier applies when the telepathist initiates the sharing process of a telepathic intuitive impression or a hallucination with a participant of which seems to occur irregularly and of which the duration of the experience is momentary. An episodic hallucination may involve a quick flash of an image or an auditable single word or short phrase with the duration of the experiencing lasting only a maximum of a couple of seconds. An episodic hallucination may also involve a more "movie-like" or dynamic image or auditable whole sentences or rhymes (e.g. songs) with the duration typically lasting no longer than a few seconds. This specifier also applies when a telepathist irregularly shares an emotional state to promote adaptive or directive behavior in a participant (i.e. output), or in the telepathist (i.e. input), or both.

C. **Continuous.** This specifier applies when the telepathist initiates the sharing process of a telepathic intuitive impression or a hallucination with a participant, which seems to occur in a continual manner. In addition, this specifier applies when episodes are so frequent it is difficult for the telepathist and/or a participant to determine where one episode ends and another begins (e.g. prolonged and closely spaced episodes).

Modality Specifiers

These specifiers are for identifying the characteristic mode(s) of a telepathic simulative experience. In any case, some emotional investment in the participant, or the situation in which the participant resides, on the telepathist's behalf is expected.

A. **Dream.** Refers to telepathic simulation during sleep where the telepathist initiates the sharing processes of information with a participant during the dream state to promote – once awake -- adaptive or directive behavior in either the telepathist or the participant.

D. **Intuitive Impressions/Emotional.** Refers to non-hallucinatory sensations of which can be described as telepathic emotional content shared between the telepathist and a participant that results in adaptive or directive behavior in the participant (i.e. output), or in the telepathist (i.e. input), or both.

B. **Auditory Hallucinations.** Hallucinations of hearing/sound. Typically only involves <u>verbal</u> hallucinations as opposed to non-verbal hallucinations. While the origin of telepathic auditory hallucinations are external, they are typically perceived as <u>internal</u> (i.e. heard within the mind as opposed to seemingly heard by the physical ear), and due to the non-invasive nature of the hallucination, auditory hallucinations are assumed by the telepathist and participants to be an auditory thought of their own volition that has spontaneously come to mind. For example, the telepathist and a participant suddenly begin thinking about the same word or phrase, rather than a telepathist thinking about a subject thinking about a word or phrase (TC), or a telepathist attempting to impress a subject to think about a word or phrase (TI).

C. **Visual Hallucinations.** Hallucinations of sight. Involving a perceived complexity classified as <u>simple</u> or <u>complex</u>. Common visual hallucinations are those located beyond the visual field (e.g. in the back of the mind, third eye vision, etc.) classified as <u>extracampine</u> hallucinations. Using the perceived shape of the hallucination, visual hallucinations can be classifies as <u>formed</u>, <u>organized</u>, or <u>unformed</u> (i.e. abstract). Telepathic simulative visual hallucinations do not take over the telepathist's or a participant's visual field. Rather, they are non-intrusive and often go unnoticed by both the telepathist and the participant (i.e. consciously unseen/subconsciously viewed, or consciously viewed with little to no knowledge that image is "shared."). For example, the telepathist and a participant suddenly begin thinking about the same image, rather than a telepathist thinking about a subject thinking about an image (TC), or a telepathist attempting to impress a subject to think about an image (TI).

D. **Tactile Hallucinations.** Hallucinations of pressure and touch. Can include a wide range of sensations from a pat on the shoulder, a knee injury, a blow to the head, and hot and cold sensations. Tactile hallucinations are classified based on the type of sensation experience (e.g. painful sensations are classified as pain hallucinations; temperature sensations are classified as thermal/thermic hallucinations). Tactile hallucinations, especially pain hallucinations, are typically a telepathist-to-participant or a participant-to-telepathist simulation of a tactile sensation with identical or nearly identical location and intensity of sensation across all involved in the simulative process. In other words, the location (e.g. right knee) and intensity of the sensation is mimicked across all involved in the simulative process and those involved typically assume the sensation is entirely natural (non-synthetic). This is less common in subjects of telepathic interaction, where the telepathist does not need not feel a sensation personally to impress a sensation onto a subject, and where the subject is often capable of identifying the telepathist or at least identifying that the sensation is synthetic.

E. **Somatic Hallucinations.** Hallucinations from inside the body (e.g. heart, lungs, sensations within the limbs, stomach e.g. nausea). Also known as somatosensory hallucinations. These hallucinations are typically a telepathist-to-participant or a participant-to-telepathist simulation of a somatic sensation with identical or nearly identical location and intensity of sensation across all involved in the simulative process, and are sensed as non-synthetic.

F. **Olfactory Hallucinations.** Hallucinations of smell. These hallucinations are typically intrinsic where the smell is typically assumed by the non-initiating participant to be the result of a spontaneous triggered or subconsciously triggered olfactive memory (e.g. the smell of tobacco, fumes from a fire, flowers and grass in a park, the perfume of a loved one, etc.). In this case, olfactive hallucinations are typically a telepathist-to-participant or a participant-to-telepathist simulation of a smell with identical or nearly identical odor and intensity across all involved in the simulative process, and are sensed as non-synthetic.

G. **Gustatory Hallucinations.** Hallucinations of taste. May include a wide range of taste sensations classified as <u>bitter</u>, <u>sour</u>, <u>sweet</u>, "<u>disgusting</u>," etc., but can be classified in more specific terms (e.g. tobacco, garlic, salt, blood, etc.). These hallucinations are typically <u>intrinsic</u> where the taste is often assumed by the telepathist and/or participant to be the result of a spontaneous triggered (i.e. subconsciously triggered) gustatory memory. In this case, gustatory hallucinations are typically a telepathist-to-participant or a participant-to-telepathist simulation of a taste with identical or nearly identical flavor and intensity across all involved in the simulative process, and are sensed as non-synthetic.

H. **Compound.** Several modalities are involved, in which case each mode involved should be noted.

Associated Mental Health Findings

Mental health disorders somewhat common in experients of telepathic simulation include: Alcohol and/or Substance Abuse/Dependence; Attention Deficit/ Hyperactivity Disorder; Bipolar Disorder; Depressive Disorder; Generalized Anxiety Disorder; Obsessive Compulsive Disorder; Panic Disorder with or without Agoraphobia; and Social Phobia (Kelly, 2011a).

Associated Medical Condition Findings

Physical medical conditions somewhat common in experients of telepathic simulation can include: Acne; Asthma; Cancer (e.g. brain, lung, breast, etc.); Kidney Diseases; Migraines; Multiple Sclerosis; Prostate Conditions; Tinnitus; Tonsillitis (Kelly, 2011a).

Differential Classification

A wide variety of extrasensory phenomena can present with similar phenomenology. These include:

o **Empathy.** Applied when there is evidence to support that emotional content is the only type of content perceived by the telepathist and a participant. However, if other informational content is involved, the experience should be classified as telepathy.

o **Telepathic Cognition.** Applied when there is evidence to support that the telepathist is only capable of one-way, subject-to-telepathist communication in the form of dual independent thought and the acquisition of knowledge.

o **Telepathic Interaction.** Applied when there is evidence to support that the telepathist is only capable of one-way, telepathist-to-subject communication in the form of dual independent thought and impression.

o **Clairvoyance.** Applied when there is evidence to support that the information obtained was 'about' an individual but the information obtained is not 'from' the individual (i.e. is indirect). Information received via telepathic simulation is typically in first-person plural (i.e. "We" feel or "We" think) from either individual's perspective; while information 'about' an individual typically results in third-person information (i.e. "She" feels or "He" thinks). If an individual becomes aware of an ailment in another individuals body, but no other individual was aware of the physical ailment, then this would be classified as clairvoyance. This is because telepathy is mind-to-mind communication, not mind-to-body communication, and telepathy must include at least two individuals, and because the knowledge of the ailment did not originate from another mind.

Criteria for Telepathic Simulative Experiences

A. **Characteristic phenomenology:** all of the following are required criteria for telepathic simulative experiences including criteria for telepathy in general.

   (1) Information is shared between the telepathist and one or more participants.

   (2) Information is shared in first-person plural perspective (e.g. If visual: the image is shared with the telepathist and all participants, and is from a group perspective involving all other participants) or narrative (e.g. If auditory: the words shared are in first-person plural perspective i.e. "We want to behave this way.").

   (3) Subconscious need for information sharing present at the time of the experience.

2. CLAIRVOYANCE

Mind-to-Environment, Object, or Entity Communication

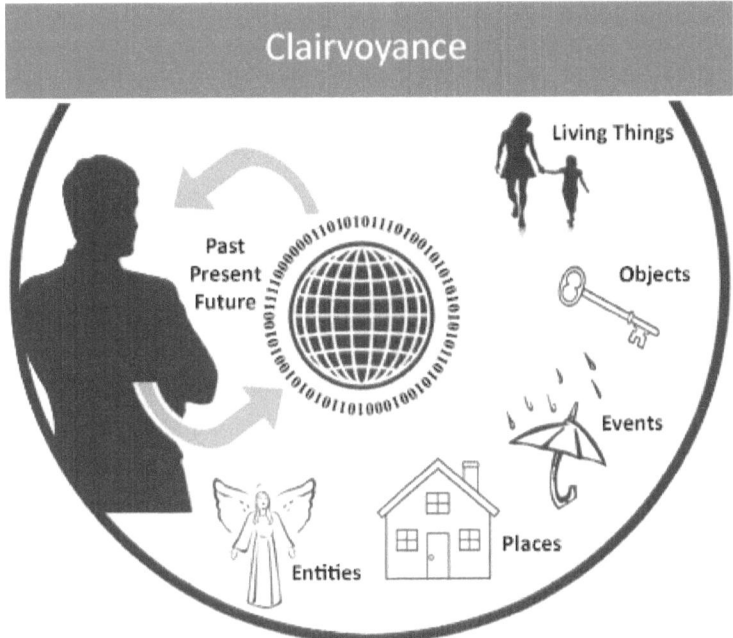

Phenomenological Features

The psychic phenomena in this section include Generalized Clairvoyance, Clairvoyant Cognition, Clairvoyant Interaction, and Clairvoyant Simulation. These phenomena have been grouped together to facilitate differential classification of phenomena that include Clairvoyance as a prominent aspect of the experience. The term *clairvoyance* has historically received numerous definitions, none of which is universally accepted across both the scientific and religious/spiritual communities. The shortest scientific definition of clairvoyance is "a form of extrasensory perception." A slightly more detailed definition describes clairvoyance as "the phenomenologically indirect knowledge of another person's thoughts or mental states" (Braude, 1978), or historically as "the knowledge of objects or states gained without the use of the known senses" (Myers, 1903). In more detail, clairvoyance is defined as a "paranormal acquisition of information concerning an object or contemporary physical event; in contrast to telepathy, the information is assumed to derive directly from an external physical source (such as a concealed photograph), and not from the mind of another person" (Thalbourne, 2003).

In this section, clairvoyance is defined as the "psychical influence of a hypothetical objective environment, universal information system, or Nature, which is assumed capable of storing, retaining, and recalling information pertaining to the past and current states of people, objects, and events; involved in the interim integration, processing, shifting, and retrieval of information pertaining to people, objects, and events in real-time, and probabilistically determining the potential trajectory of future events" (Kelly, 2011b). Clairvoyance is presumed possible through the act of an experient requesting and receiving information pertaining to past events via the systems "long-term information storage," or pertaining to real-time events via the systems "working memory" or "short-term information storage."

Information pertaining to future events may be the result of Natures computational capabilities of relative causal knowledge encompassing deterministic and random variables, which may be stored by Nature and retrievable by the experient. Nature appears to capture, retain, and store information, and this information can be requested by an experient in which is then conveyed intuitively and in the form of sensory hallucinations, primarily through the visual and auditory modalities. These hallucinations can also occur in other sensory modalities including olfaction, gustatory, and somatosensory.

In addition, clairvoyance includes "anomalous communication with immaterial entities most commonly referred to as apparitions (e.g. ghosts) or spirit guides, which are believed to have a form of consciousness and element of personality" (Kelly, 2001b). Assuming thought to be the basis of consciousness, this form of communication is assumed to require specific mental processes, which experients utilize for non-local thought transference between the experient and an immaterial entity. These thoughts appear to be transferred through intuitive, or emotional, modes or through several hallucinatory sensory modes including visual, auditory, olfaction, gustatory, and somatosensory modalities. (Kelly, 2011b).

Subtypes

The following subtypes are phenomenological subgroups exclusive to clairvoyance only.

2.1 (CC)   **Clairvoyant Cognition** (see p. 59)

2.2 (CI)   **Clairvoyant Interaction** (see. p. 69)

2.3 (CS)   **Clairvoyant Simulation** (see p. 78)

Severity Specifiers

These specifiers should only be used when all criteria for the type of a subtype are currently met. In deciding whether reported experiences should be described as stable/functional, mild, moderate, or severe, the clinician should take into account the number and intensity of the experiences and any resulting impairment in occupational or social functioning.

A. **Stable/Functional.** Intentional experiences of which fit all criteria with few, if any, spontaneous experiences and of which result in no impairment in social or occupational functioning and may or may not increase normal functioning.

B. **Mild.** Few experiences of which fit all criteria and experiences result in no more than minor impairment in social or occupational functioning.

C. **Moderate.** Experiences and functional impairment between "mild" and "severe" are present.

D. **Severe.** Many experiences of which fit all criteria, either episodic or continuous, of which result in marked impairment in social or occupational functioning.

Associated Research and Laboratory Findings

No laboratory findings have been identified that are diagnostic of clairvoyance. However, a variety of measures from neuroimaging, neuropsychological, and neurophysiological studies have shown differences between groups of individuals with clairvoyance and appropriately matched control subjects. According to Williams & Roll (2000), in studies examining the correlation between clairvoyant scoring and alpha abundance, positive correlations have been found except in precognitive designs, where negative correlations have been found. In one study examining correlations between clairvoyant scoring and changes in alpha frequency and amplitude, positive correlations have been found, but further independent replication is needed. In studies examining the correlation clairvoyant scoring and the effects of alpha feedback training, negative correlations have been found. There have been several studies examining the correlation between clairvoyant scoring and other types of brain waves (e.g. gamma and beta), but further studies with broad-spectrum EEG are required to determine which types to what degree.

In experimental studies examining the precognitive response at an unconscious level (i.e. presentiment), changes in brain activity preceding the onset of emotional stimuli consists of voltage changes across the cortex, which were found in ERP studies, and changes in blood oxygenation in the areas involved in the processing of the associated sensory stimuli, which were found in fMRI studies. Predictions have been made that the right hemisphere may be psi-conducive while the left hemisphere may be psi-inhibitive. While the evidence for this prediction is not definitive, most relevant studies suggest that the right hemisphere is most associated, and may even be exclusively associated, with clairvoyant experience. EEG studies on two notable psychics suggest right hemisphere processing, the right medial and superior parietal lobes, but additional brain wave measurement and imaging studies need to be conducted with other notable psychics to make any further determinations.

Numerous studies have implicated the temporal lobe as the region that shapes extrasensory experience. One study has found that individuals with temporal lobe dysfunction reported more extrasensory (psi experiences in general) experiences than other patients. Three studies involving mediums and psychics found elevated temporal lobe signs. Predictions have been made that the hippocampus and amygdala are activated during extrasensory experiences because (1) numerous studies have indicated that extrasensory response consists of implicit emotional memories in which correspond to a perceived object, and (2) memory and emotion are processed by these regions. Another prediction is that precognition and retrocognition may activate the medial temporal regions of the brain. Further brain regions that may be associated with extrasensory experience are the occipital lobe and the parietal lobe.

A 13 year experiment by Kolodziejzyk (2012), utilizing a unique approach to associative remote viewing (ARV) with a total of 5,677 ARV trials, yielded a statistically significant score. Most of the project questions utilized focused on predicting the outcome of a given futures market, making this a remote viewing/precognitive experimental design. According to Radin (2006), in two experiments investigating EEG correlations in separated pairs of individuals utilizing a protocol of photic stimulation and EEG measurements, one of which involved two identical twins, followed by 10 replications, 8 of the studies were reported positive. Many replications followed over the years, with one team concluding that the phenomenon could not be easily dismissed and no biophysical mechanism known could account for the correlations.

A further replication, where the subject was placed in an fMRI scanner and the agent in a distant room, they found a highly significant increase in brain activity in the subject's visual cortex while the agent was viewing a flickering light. However, while the experiment's design was intended to be telepathic, a pre-stimulus or "presponse" (i.e. physiological activity before the stimulus) has been found in the subject during experiments suggesting a telepathy/clairvoyance (presentiment) design. This "presponse" can also be found in experiments by Radin (2000), where skin conductance changes before, during, and after the presentation of randomly selected emotional and calm pictures, where larger average arousal levels are found during the period before the display of emotional pictures as compared to before calm pictures.

In a study by McCraty et al. (2004), and in two different experiments conducted by Sartori, Massaccesi, Martinelli, & Tressoldi (2004), there were reports of presentiment in experiments utilizing skin-conductance, heart rate, and EEG measures. The former experiment was positive in regard to skin-conductance, though not significant, and found that heart rate significantly slowed prior to an emotional stimulus, that women performed better than men, and that the brain responded in a different manner prior to emotional and calm stimuli. In the latter two experiments, rather than emotionally evoking pictures, this method involved the presentation of targets and non-targets. In the first experiment (general clairvoyant design) and second experiment (precognition/presentiment design), results were significant where heart rate associated with targets increased significantly compared to non-targets, but the mean of correct hits was close to chance expectation.

According to Radin (2000), in another type of precognitive/presentiment experiment, involving reaction time and contingent negative variation (CNV), a slow brainwave indicator of anticipation was utilized to unconsciously detect a stimulus that would randomly appear in the future. The results were highly significant just before the target stimulus appeared. In a triple-blind study by Beischel & Schwartz (2007), examining the anomalous reception of information about deceased individuals by research mediums, results were highly significant compared to control. According to Spottiswoode (1990), laboratory experiments involving a remote viewing design revealed a negative correlation between scores in contemporaneous clairvoyant perception and geomagnetic fluctuations (i.e. that scores were higher during times when the electromagnetic activity of the earth was lowest. This negative correlation was not found for precognition or retro/postcognition. Variations in the GMF field were indicated by the *ap* geomagnetic index.

Specific Culture, Age, and Gender Features

Clinicians assessing beliefs and claims in socioeconomic or cultural situations that are dissimilar from their own must take cultural dissimilarities into account. Ideas that may appear to be questionable or even delusional in one culture or subculture (e.g. Buddhists, New Agers, Spiritualists, Wiccan Practitioners, and those who engage in regular meditative practices) may be commonly believed in another. In some cultures, clairvoyant hallucinations with a spiritual or religious content may be a normal part of spiritual or religious experience (e.g. the sensing of a loved one after they have passed on, mothers intuition; especially in a precognitive context where a foreknowledge of danger is perceived; or the emotional arousal associated with experiencing the Holy Spirit). These varying beliefs may have subtle to blatant differences in terminology and descriptions leaving the clinician with the difficult task of properly categorizing experiences into parapsychological types and subtypes.

In regard to physical location, in a study conducted by Haraldsson & Houtkooper (1991), individuals in the U.S. reported clairvoyance experiences 25%, and contact with the dead 30%. In European nations surveyed, with clairvoyance experiences listed first and contact with the dead listed second, Italy reported 39% | 34%, France reported 24% | 24%, West Germany reported 17% | 28%, and Finland reported 15% | 14%, with Great Britain, Spain, and Belgium each reporting 14% for clairvoyant experiences and in order of listing 26%, 16%, 18% for contact with the dead. The lowest percentages were found in Sweden, Norway, and Iceland with only 7% reporting clairvoyant experiences each, with in order of listing 14%, 9%, 41% reporting contact with the dead, putting Iceland in the lead for the higher percentage of reporting contact with the dead.

Initial experiences (onset) of clairvoyant phenomena typically occur within the first several years after birth and/or during puberty. Early onset may involve several spontaneous experiences of which may or may not affect the child psychologically, emotionally, or socially. Experiences in which have an early onset and continue throughout life without extended pause (e.g. one year or more without an experience) typically remain stable/functional in the long term. In some generalized ESP experiments, children tend to score higher than adolescents and adults. However, many similar studies have been unsuccessful in in demonstrating age dependent differences in scoring (Palmer, 1978). According to Blackmore (1980), "Although many studies show high scoring in children there is little systematic evidence of a relationship between ESP and age and there are many contradictory findings."

The onset of clairvoyant phenomena during puberty, most common between the ages of 13-16, is typically induced to compensate for an inability to effectively communicate their wants, needs, and/or thoughts verbally to other individuals, or lack individuals in their life that could properly meet their needs. Experients may feel they have had a recent decline in quality of life, academic performance, and/or social relationships. During this time experiences are typically spontaneous, and can range from mild to severe depending on the severity of needs the experient feels they are unable to communicate and obtain. Experiences in which are moderate to severe have a high probability of continuing in severity unless the want or needs of the adolescent are met.

Onset during this age rage may also be induced by another individual unwilling to meet the experients wants or needs, or the adolescent's general environment may be unaccommodating in some manner. In other words, the experient is communicating effectively, but the recipient does not understand effectively (e.g. a parent that does not understand a child's limitations due to physical or mental illness), or simply refuses to meet the experient's needs (e.g. a bully at school, or an abusive or neglectful parental figure), or the experient is able to communicate effectively, but the resources required are not being made available (e.g. food, shelter, clothing, etc.)

Adult onset may occur at any age and is typically precipitated by and inability to verbally communicate wants, needs, or thoughts, physically acquirer wants or needs, possess a desire to continue a connection with a deceased individual, deepen a mental and emotional connection with another individual(s) (e.g. an individual in the experient's life or a type of individual the experient wants in their life). Adult onset is typically stable/functional to mild unless precipitated by experiences that amount to trauma, illness, or any other type of sudden uncomplimentary experience, acute or chronic, that results in a major disturbance in the experient's life. In the case of the latter, moderate to severe experiences are typically common. Spontaneous experiences are common regardless of the severity. However, stable/functional to mild experiences are more likely to be the product of intention, while moderate to severe experiences are mainly spontaneous.

Gender differences have been the focus of some studies. Overall, there appears to be no clear trend for differential scoring between males and females (Palmer, 1978), and if gender differences are found, they tend to be slight with women reporting clairvoyant experiences more than men, men reporting clairvoyant experience more than women, or no difference between gender reporting was found. However, strong differences have been found in reports of contact with the dead In Europe, 30% of women and only 20% of men report this type of communication, while in the U.S., the difference is 34% for women and 25% for men (Haraldsson & Houtkooper, 1991).

Familial Patterns

Occasionally one biological parent or grandparent of an experient of clairvoyant phenomena reports a history of clairvoyant-like experiences. Familial patterns most common are clairvoyant experiences between mother and child, spouses/lovers, identical twins, and occasionally between fraternal twins, siblings, and meditation partners. In regard to marital status in Europe and the U.S., relatively fewer single and married individuals report contact with the dead then the "combined broken relationship group" (i.e. living as married, separated, divorced, or widowed), with clairvoyant experiences having a similar effect, but to a lesser extent (Haraldsson & Houtkooper, 1991).

Associated Terminology

Emotional content. Experients of clairvoyance in which primarily sense emotional content, but still receive more than emotional content on occasion, may use the following terminology: clair-empathic • clair-empathy • empath • empathic • empathist • empathy • gut feeling • intuition • intuitionism • intuitive • intuitvism • keen intuition • presentiment • mothers intuition • sensitive.

Visual content. Experients of clairvoyance in which primarily sense visual content may use the following terminology: auras • clairvoyant • clairvoyance • medical intuition • medium • mediumship • precognition • precognitive • postcognition • postcognitive • psipath • psipathic • psychic • remote viewer • remote viewing • retrocognitive • retrocognition • premonition • psychometric • psychometry.

Additional terminology. Used in a context involving the experient and a person, object, discarnate, entity, or location/environment, including both clairvoyance and clairvoyance-like terminology: aha! moment • angel • anomalous communication • audible thoughts • augar • augury • channeller • channeling • deceased • discarnate • distance healing • divination • diviner • eureka effect • fortune teller • guardian angel • ghost • ghost whisperer • haruspex • holy spirit • insight • intention • intention healing • magic • magick • miracle • oracle • palm reader • prayer • prayer fulfillment • prayer healing • predictor • probability shifting • prophesier • prophet • psychic communication • psychic knowledge • second sight • seer • shaman • shamanic • sibyl • six sense • soothsayer • spell casting • spirit • spirit guide • spiritualist • visionary • wish fulfillment.

Criteria for Clairvoyant Experiences

A. **Characteristic phenomenology:** all of the following are required criteria for clairvoyance.

   (1) Mind-to-environment/nature (including information about a person), mind-to-object, or mind-to-entity communication/effect.

   (2) Involves one or more environments, objects, entities, or indirect (about) information pertaining to a person's situation.

   (3) The source of the information is not a living organism (i.e. the source is the environment/nature, a discarnate entity, or other entity), but information can be obtained, or probability influenced, pertaining to a living organism's situation (e.g. health, environment, current events).

B. **Social/occupational need:** A subconscious need has been identified as the catalyst for the initiation of clairvoyant processes; i.e. (1) identified an inability to communicate wants, needs, or thoughts to an individual in an interpersonal, academic, or occupational context, (2) inability to acquire physical necessities, (3) desire to continue a connection between themselves and a discarnate entity, or (4) need for information not readily accessible though natural means, etc.

C. **Validation:** The experience has been validated by an individual other than the experient (e.g. a sitter [an individual in which asks a psychic medium to contact a discarnate entity on their behalf reports accuracy in intermediated information], or news report, etc.), and the clinician determines the experience was more than a coincidence/chance occurrence based on the quality of the information received and reported, and all other possible explanations for obtaining the information is excluded. If validation does not apply, yet clairvoyant processes are still plausible, the experience should be classified as "Possible Clairvoyance" (PC).

D. **Empathy Exclusion:** Psychical empathy has been ruled out because more than emotional content is involved in the experience(s).

E. **Telepathy Exclusion:** Telepathy has been ruled out because mind-to-mind communication is not the basis of the experience(s).

Clairvoyance Subtypes

The subtypes of clairvoyance are defined by the predominant phenomenology of reports. The determination of a particular subtype is based on the clinical picture that occasioned the most recent experiences, and may therefore change over time. Not infrequently, the description of experiences may include phenomena that are characteristic of more than one subtype. The choice among subtypes depends on the following algorithm: Clairvoyant Cognition (CC) is assigned whenever information is clairvoyantly acquired by the experient originating from an environment, object, entity or about an individual; Clairvoyant Interaction (CI) is assigned whenever information is clairvoyantly acquired/conveyed through a medium in a trance state by an entity and of which involves automatism(s) to some degree; Clairvoyant Simulation (CS) is assigned whenever an accommodating/shifting effect is initiated by the experient or the experient's environment; If two or three subtypes are assigned, all should be listed; Generalized Clairvoyance (GC) is assigned when all subtypes appear to apply (optional), or a clear choice is unable to be made, but appears to only suggest clairvoyant phenomena. In addition, when a clear choice cannot be made, the clinician should consider a dimensional approach to classifying the experiences.

2.1 Clairvoyant Cognition (CC)

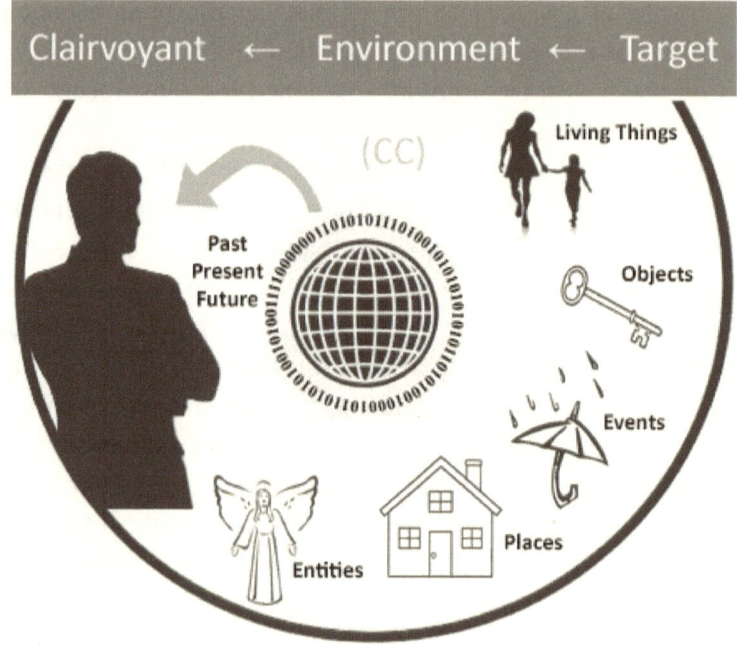

Clairvoyant ← Environment ← Target

(CC)

Past
Present
Future

Living Things

Objects

Events

Places

Entities

Phenomenological Features

The essential feature of the (CC) type of clairvoyance is the *phenomenologically indirect knowledge of an individual, object, or event via Nature.* In cases of clairvoyant cognition, an experient is retrieving information from Nature, i.e. the experient is able to "pick up on" the information recorded in Nature. The informational system from which the information originates does appear to play an intentional part in the information teleportation processes. In other words, the process is not assumed entirely evoked by the receiver, whereby possibly suggesting Natures ever-present contribution to the survival (well-being) of the experient. Again, in regard to clairvoyant cognition, the experient is an individual capable of evoking clairvoyant processes, or allowing the occasion for Nature to evoke such processes. Here, the experient will become aware of Nature-based information (e.g. states), but should be able to clearly identify that the information did not originate in their own mind. Here the information is received and perceived by the experient, but the information did not develop from a chain of prior thoughts belonging to the experient. Instead, the information appears to "pop up," but is immediately associated with a specific event, object or individual other than the experient, or simply identified as not originating from the experient (Kelly, 2011b).

Mental mediumship is defined as *the anomalous communication with immaterial entities* most commonly referred to as discarnate spirits (i.e. ghosts) or spirit guides, which are believed to have a form of consciousness and element of personality (Irwin & Watt, 2007), or the anomalous communication with Nature, which some believe to be an aggregate of consciousness, or "universal consciousness," assumed "friendly," (Braud, 2011), but possessing a neutral personality (i.e. behaviors, temperament, emotions, etc.). However, anomalous communication with discarnate spirits is more commonly associated with mediumship, while the anomalous communication with Nature is more commonly associated with clairvoyance, or clairvoyant cognition. Mental mediumship typically involves a medium communicating with, or receiving information pertaining to a discarnate spirit. However, other types of entity communications such as "demons" or "angels" are also commonly reported (Kelly, 2011b).

Mental mediumship is a mode of clairvoyant cognition, in that the experient is communicating with Nature or an entity that is not "physically" present. Information is transferred through intuitive, or emotional, modes or through several hallucinatory sensory modes including visual, auditory, olfaction, gustatory, and somatosensory modalities. The most common sensory modalities utilized by mediums are the audio (clairaudient) and visual (clairvoyant) modes with the occasional exception of sensory modes associated with clairsentience. Mental mediums tend to report having an open connection with Nature, or entities, or employ a spirit guide to relay information back and forth to other entities (Kelly, 2011b).

Intention Specifiers

The first set of specifiers is for identifying whether the experience was intentional or unintentional.

A. **Spontaneous.** This specifier applies when information appears to "pop into mind" rather than being intentionally requested by the percipient.

B. **Intentional.** This specifier applies when a percipient selects or specifies the type or source of information required or entity to contact (discarnate or other).

The second set of specifiers is for identifying the subconscious or conscious need or goal that is assumed to be the catalyst for initiating clairvoyant cognitive processes.

A. **Adaptive.** This specifier applies when information acquisition is initiated to assist the percipient in understanding and/or adapting to an individual, object, or event to which they typically have some level of emotional investment. This specifier also applies when the experient perceives information in order to assist another in understanding and/or adapting (e.g. a sitter in regard to the loss of a loved one, a discarnate in regard to his or her own passing, etc.).

B. **Decisive.** This specifier applies when information acquisition is initiated to assist the percipient in coming to a decision involving an individual, object, or event in which they typically have some level of emotional investment. This specifier also applies when the experient perceives information in order to assist another in coming to a decision.

The third set of specifiers is for identifying the temporal feature of the clairvoyant cognitive experience.

A. **Precognition.** This specifier applies when information acquisition pertains to the potential trajectory of a future state of an individual, object, or event. In other words, when probabilistic information is perceived by an experient. This specifier includes <u>presentiment</u>

and <u>premonitions</u> (i.e. where information perceived in regard to future events involves only emotional content).

B. **Contemporaneous.** This specifier applies when information acquisition pertains to the present/current state of a local, distant, or unseen individual, object, or event (i.e. information pertaining to real-time states).

C. **Retro/Postcognition.** This specifier applies when information acquisition pertains to the past state of an individual, object, or event (i.e. information pertaining to historical states that may or may not be recent, but are not current).

The fourth set of intention specifiers is for identifying the source of the information acquired.

A. **Nature.** This specifier applies when information acquired is not directly from a living individual (e.g. human or animal), not from a discarnate, and not from some other described single entity.

B. **Discarnate.** This specifier applies when information is acquired from a once living but now deceased individual (e.g. human or animal).

C. **Other Entity.** This specifier applies when information is acquired from an entity that has never lived in the physical sense (e.g. human or animal) but is described as living non-physically and possessing a single consciousness (e.g. angels or spirit guides that never lived as human).

The fifth set of intention specifiers is for identifying what the information perceived pertains too (i.e. the target).

A. **Individual.** This specifier applies when information is acquired pertaining to the experient or another individual(s) (i.e. person, animal, discarnate, entity), such as features of that individual (e.g. the individuals location, viewpoint, eye color, health, etc.).

B. **Object.** This specifier applies when information is acquired pertaining to an inanimate physical object(s) (e.g. a watch, necklace, home, etc.), such as features of that object (e.g. the objects current location, size, color, position, previous owner, etc.).

C. **Event.** This specifier applies when information is acquired pertaining to a situation (e.g. a birth, death, meeting, divorce, crime, wedding, car accident, etc.), such as features of that event (e.g. effect, consequence, issue, outcome, probability, result, state of affairs, etc.)

Development and Course

Childhood onset may present itself through dreams, visual and/or auditory hallucinations, with intuitive impressions (i.e. gut feelings, intuition) being also common. Adolescent onset primarily presents itself through visual and/or auditory hallucinations with clairvoyant dreams and intuitive impressions being also common. However, other types of hallucinations (e.g. olfactory, tactile, etc.) are less common. Adult onset primarily presents itself through clairvoyant dreams, intuitive impressions, or during crisis situations, in the form of hallucinations subconsciously deem most appropriate for notification.

Course Specifiers

These specifiers are for identifying the characteristic course of clairvoyant cognitive experiences over time.

A. **Single Episode.** This specifier applies when the percipient experiences a clairvoyant intuitive impression or hallucination without a prior history of episodes.

B. **Episodic.** This specifier applies when the percipient experiences clairvoyant intuitive impressions or hallucinations of which seem to occur irregularly and of which the duration of the experience is very momentary. An episodic hallucination may involve a quick flash of an image or an auditable single word or short phrase with the duration of the experiencing lasting only a maximum of a couple of seconds. An episodic hallucination may also involve a more "movie-like" or dynamic image or auditable whole sentences or rhymes (e.g. songs) with the duration typically lasting no longer than a few seconds. While the percipient may appear distracted during a clairvoyant cognitive episode, the experient should still be fully aware of their surroundings.

C. **Continuous.** This specifier applies when the percipient experiences clairvoyant intuitive impressions or hallucinations of which seem to occur in a continual manner, or when episodes are so frequent it is difficult for the percipient to determine where one episode ends and another begins (e.g. prolonged and closely spaced episodes).

Modality Specifiers

These specifiers are for identifying the characteristic mode(s) of a clairvoyant cognitive experience. In any case, some emotional investment in the individual, object, or the situation on the experients, or sitters, behalf is expected.

A. **Dream.** Refers to clairvoyant information acquisition during sleep.

B. **Intuitive Impressions/Emotional.** Refers to non-hallucinatory sensations of which can be described as clairvoyantly received emotional content.

C. **Auditory Hallucinations.** Hallucinations of hearing/sound. Typically only involves verbal hallucinations as opposed to non-verbal hallucinations. While the origin of clairvoyant auditory hallucinations are external, they are typically perceived as internal (i.e. heard within the mind as opposed to seemingly heard by the physical ear), and stem from an often identifiable source (e.g. Nature, discarnate, other entity).

D. **Visual Hallucinations.** Hallucinations of sight. Involving a perceived complexity classified as simple or complex. If the entire environment is replaced by the visual hallucination, the hallucination is classified as scenic or panoramic hallucinations. Visual hallucinations in which are located beyond the visual field (e.g. in the back of the mind, third eye vision, etc.) are classified as extracampine hallucinations. Using the perceived shape of the hallucination, visual hallucinations can be classified as formed, organized, or unformed (i.e. abstract).

E. **Tactile Hallucinations.** Hallucinations of pressure and touch. Can include a wide range of sensations from a pat on the shoulder, a knee injury, a blow to the head, or hot and cold sensations. Tactile hallucinations are classified based on the type of sensation experience (e.g. painful sensations are classified as pain hallucinations; temperature sensations are classified as thermal/thermic hallucinations).

F. **Somatic Hallucinations.** Hallucinations from inside the body (e.g. heart, lungs, sensations within the limbs, stomach e.g. nausea). Also known as somatosensory hallucinations.

G. **Olfactory Hallucinations.** Hallucinations of smell. These hallucinations are typically extrinsic where the localization of the smell is outside of the body (e.g. the smell of tobacco, fumes from a fire, flowers and grass in a park, the perfume of a loved one, etc.).

H. **Gustatory Hallucinations.** Hallucinations of taste. May include a wide range of taste sensations classified as bitter, sour, sweet, "disgusting," etc., but can be classified in more specific terms (e.g. tobacco, garlic, salt, blood, etc.).

I. **Compound.** Several modalities are involved, in which case each mode involved should be noted.

Associated Mental Health Findings

Mental health disorders somewhat common in experients of clairvoyant cognition include: Alcohol and/or Substance Abuse/Dependence; Attention Deficit/ Hyperactivity Disorder; Depressive Disorder; Generalized Anxiety Disorder; Obsessive Compulsive Disorder; Panic Disorder with or without Agoraphobia, and Sleep Disorder (Kelly, 2011b).

Associated Medical Condition Findings

Physical medical conditions somewhat common in experients of clairvoyant cognition can include: Chronic Fatigue Syndrome, Chronic Pain and Neurological Disorders (Asperger's Syndrome, Autism, Myalgia, Electromagnetic Hypersensitivity, Fibromyalgia, Multiple Chemical Sensitivity, Myofascial Pain Syndrome, Sensory Processing Disorder, Temporomandibular Joint Disorder, etc.), Diabetes, Digestive Disorders, Food Allergies/Sensitivities, Hyperthyroidism, and Hypoglycemia (Kelly 2011b).

Differential Classification

A wide variety of extrasensory phenomena can present with similar phenomenology. These include:

o **Empathy**. Applied when there is evidence to support that emotional content is the only type of content perceived by the percipient. However, if other informational content is involved, the experience should be classified as clairvoyance.

o **Psychometry.** Applied when clairvoyant cognitive experiences are limited to the experient obtaining information pertaining to objects.

o **Clairvoyant Interaction.** Applied when there is evidence to support that the percipient appears to be occupied by another entity's consciousness or subject to controlled behavior (e.g. automatism, xenoglossy), discarnate or otherwise.

o **Clairvoyant Interaction**. Applied when there is evidence to support physical mediumship or psychopompic activity.

o **Clairvoyant Simulation.** Applied when there is evidence to support that the percipient is a participant in regard to information acquisition or shifting probability to create an accommodating effect.

o **Telepathy**. Applied when there is evidence to support that information is obtained directly from a living organism (e.g. human or animal), rather than obtaining information about an individual from Nature, a discarnate, or other type of single consciousness entity. Information received telepathically is typically in first-person, second-person, or "direct," while information received clairvoyantly is typically in third-person or "indirect." (e.g. if the percipient receives information described "as though they are looking through the eyes of another individual," this would be classified as telepathy. However, if the percipient describes receiving the information "as though they are looking at the individual and the individual's surroundings," this would be classified as clairvoyance. In a similar circumstance, if an individual becomes aware of an ailment in their own body, or the body of another individual, but no other individual was aware of the physical ailment, then this would be classified as clairvoyance. This is because telepathy is mind-to-mind communication, not mind-to-body communication, and telepathy must include at least two living individuals, and because the knowledge of the ailment did not originate from another mind.).

o **Mental Mediumship**. Applied when there is evidence to support that information is obtained from only a non-physically living being (i.e. discarnate), as clairvoyant cognition refers to other types of sources of information.

Criteria for Clairvoyant Cognitive Experiences

A. **Characteristic phenomenology:** all of the following are required criteria for clairvoyant cognitive experiences including criteria for clairvoyance in general.

  (1) Information is received by the percipient through mind-to-environment (including information about a person) mind-to-object, or mind-to-entity communication.

(2) Information received is in third person perspective (e.g. If visual: the image received is viewed as though the percipient is looking at an event, object, or looking at the individual within their surroundings (i.e. rather than looking through the eyes of an individual -- telepathy), or narrative (e.g. If auditory: the words received are from the sources' perspective "You will have a fortunate day," or "She misses you dearly."

(3) Subconscious need for information acquisition present at the time of the experience.

2.2. Clairvoyant Interaction (CI)

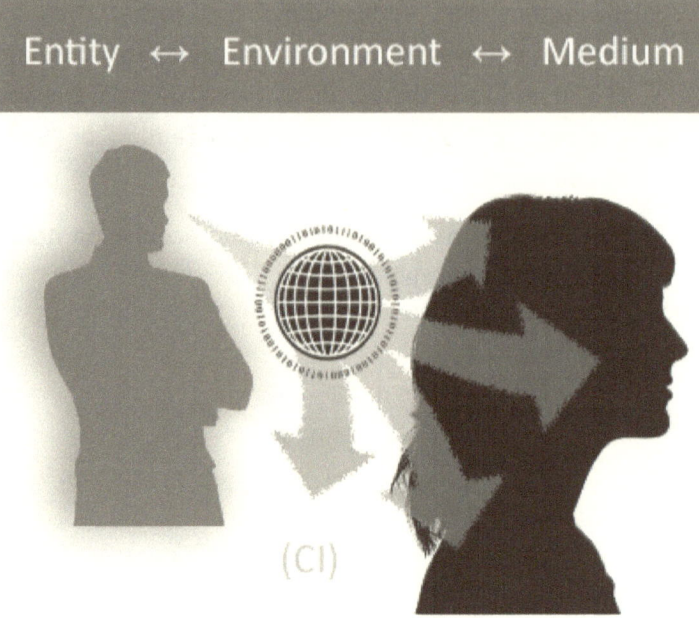

Entity ↔ Environment ↔ Medium

(CI)

Phenomenological Features

The essential feature of the (CI) type of clairvoyance is *the causal influence of an entities mind onto an experient without the intervention of the five senses.* This type of clairvoyance is typically known as trance mediumship or channeling. In either case, two forms are typically assumed to exist. The first form has characteristics of mental mediumship, but with the medium sitting or lying down in a deep meditative state. During these sessions, the medium may speak as though the information is being conveyed *to* the medium, but rather, the information is coming directly from the entity, but being conveyed *through* the medium's natural voice and behaviors.

During a session of this type, the medium's consciousness is believed to be "set aside" within their physical body as a means to allow an entity's consciousness to enter the medium's physical body. This form of mediumship, versus mental mediumship, appears to allow a means of more comprehensible communication between sitters (i.e. those desiring to speak to an entity in which are not the medium) and the entity called. Mediums in which describe "surrendering to the experience" report that they have little to no knowledge or recall of what messages where conveyed while in trance. Because of this, a second party typically assists the medium by writing down or recording (e.g. audio or audiovisual) the medium's behavior and information conveyed during the session.

The second form is characteristic of the entity influencing the medium's physical body in addition to speech (i.e. automatism or xenoglossy). The medium is often awake and aware of most of the communication period and the thoughts and words are conveyed from entity, to medium, to sitter. If the medium is aware of what is being conveyed, they typically describe the experience as though their "self" or "will" had "taken a backseat" (i.e. watching and listening, while feeling not in control of what is being said or done), but often report feeling as though they do have the ability to end the "call" at will if necessary.

In cases when an entity appears to enter a medium's physical body, where the medium often reports a feeling of surrender (i.e. unable to end the call), the medium typically reports the call lasting as long as the entity requires to fulfill their purpose (e.g. conveying information to a particular individual). Reports of clairvoyant interaction are not limited to discarnate spirits, but have also included seemingly "evil spirits" or "demonic entities, "angelic" entities, spirit guides, and omnipotent spirits described as the Holy Spirit, God(s), Goddess(es), and Nature itself. In addition, some mediums report the feeling of leaving their physical body, often triggered when the medium no longer feels entirely in control of their mental or physical faculties. While this may be suggestive of an out-of-body experience, this may simply be the result of the medium experiencing a sense of detachment rather than the medium's consciousness actually vacating their body (Kelly, 2011b).

Intention Specifiers

The first set of specifiers is for identifying whether the experience was intentional or unintentional.

A. **Spontaneous.** This specifier applies when an entity's consciousness interacts with the mind of the experient void of the experient's conscious choice, decision, or intention.

B. **Intentional.** This specifier applies when an experient selects or specifies the entity to interact with, and is open to the experience of entity interaction.

The second set of specifiers is for identifying the subconscious or conscious need or goal that is assumed to be the catalyst for initiating clairvoyant interactive processes.

A. **Adaptive.** This specifier applies when interaction is initiated to assist the experient or a sitter in understanding and/or adapting to the loss of an individual of which they typically have some level of emotional investment. However, understanding and/or adaptation may also be consciously or subconsciously required by the experient or sitter in regard to the self or a past, present, or up-and-coming situation involving some level of emotional investment on the experient's or the sitter's behalf.

B. **Decisive.** This specifier applies when interaction is initiated to assist the experient or a sitter in coming to a decision involving the self, another individual, object, or event/situation in which they typically have some level of emotional investment. This specifier also applies when the experient interacts with an entity in order to assist a sitter in coming to a decision in regard to the loss of an individual of which they typically have some level of emotional investment.

The third set of specifiers is for identifying the form of the clairvoyant interactive experience.

A. **Trance Mediumship.** This specifier applies when an interactive session has characteristics of mental mediumship, but with the experient sitting or lying down in a deep meditative state. Here the information is coming directly from the entity and is being conveyed *through* the experient's natural voice and behaviors.

B. **Channeling**. This specifier applies when an entity influences the medium's physical body in addition to speech. Here, if the experient is awake and aware of most of the communication period, they may describe the experience as though their "self" or "will" had "taken a backseat," and may or may not report the ability/sensation that they could end the call at any time.

The fourth set of intention specifiers is for identifying the source of the information conveyed.

A. **Nature.** This specifier applies when information conveyed is not directly from a living individual (e.g. human or animal), not from a discarnate, and not from some other described single entity.

B. **Discarnate.** This specifier applies when information is conveyed from a once living, but now deceased individual (e.g. human or animal).

C. **Other Entity.** This specifier applies when information is conveyed from an entity that has never lived in the physical sense (e.g. human or animal), but is described as living non-physically and possessing a single consciousness (e.g. angels or spirit guides that never lived as human).

The fifth set of intention specifiers is for identifying what the information conveyed pertains too.

A. **Individual.** This specifier applies when information is conveyed pertaining to the experient or another individual(s) (i.e. person, animal, discarnate, entity), such as features of that individual (e.g. the individual's location, viewpoint, eye color, etc.).

B. **Object.** This specifier applies when information is conveyed pertaining to an inanimate physical object(s) (e.g. a watch, necklace, home, etc.), such as features of that object (e.g. the objects current location, size, color, position, previous owner, etc.).

C. **Event.** This specifier applies when information is conveyed pertaining to a situation (e.g. a birth, death, meeting, divorce, crime, etc.), such as features of that event (e.g. effect, consequence, issue, outcome, probability, result, state of affairs, etc.)

Development and Course

Childhood onset may present itself through automatism, xenoglossy, physical mediumship, psychopompic activity, or compound. Adolescent onset primarily presents itself most commonly through automatism, physical mediumship, psychopompic activity, or compound, with other modes being less common. Adult onset primarily presents itself most commonly through automatism, physical mediumship, psychopompic activity, or compound, with other modes being less common.

Course Specifiers

These specifiers are for identifying the characteristic course of clairvoyant interactive experiences over time.

A. **Single Episode.** This specifier applies when the experient reports a clairvoyant interactive episode without a prior history of episodes.

B. **Episodic.** This specifier applies when the experient reports clairvoyant interactive episodes of which seem to occur irregularly and of which the duration of the experience is temporary (i.e. lasting under 1 hour in duration).

C. **Continuous.** This specifier applies when the experient reports a clairvoyant interactive episode of which seem to occur in a continual manner, or when episodes are so frequent it is difficult for the percipient to determine where one episode ends and another begins (e.g. prolonged and closely spaced episodes).

Modality Specifiers

These specifiers are for identifying the characteristic mode(s) of clairvoyant interactive experiences only. In any case, some emotional investment in the individual, object, or the situation on the experient or sitter's behalf is expected.

A. **Automatism.** (Input) This specifier applies when an entity takes control of the experient's verbal (i.e. automatic speaking) or motor skills (e.g. automatic writing, automatic drawing, use of a Ouija board, etc.). Here the experient may have no conscious awareness or volition over their own verbal or motor activity (i.e. automatic movement) resulting in the sensation of detachment.

B. **Xenoglossy.** (Input) This specifier applies when an entity impresses its own personality or skills onto an experient of which the experient himself/herself does not actually possess (e.g. communicates in a language or vocabulary unknown to the experient). In cases involving xenoglossy and the spirit of a discarnate entity, the personality or skills of the spirit, from when he/she was alive, are impressed onto the medium for the duration of the session. Here the experient may have no conscious awareness or volition over their own verbal or motor activity (i.e. automatic movement) resulting in the sensation of detachment.

C. **Physical Mediumship.** (Output) This specifier applies when there is evidence to support that the experient is influencing the energies and energy systems of entities. Here the experient may or may not report other psychical ability, but reports that they are a source of "power" for spirit manifestations such as loud rapping and other noises, voices, materialized objects, materialized spirit bodies, or body parts such as hands, or the occurrence of levitation. In other words, a physical medium typically provokes the communication of spatially located spirits. Here the physical manifestations are perceptible to other individuals. This modality is the product of spirit-based telekinetic phenomena where the entity is the source of the disturbances and the experient is the source of the entity's additional "power." It is debatable whether this modality is extrasensory or psychokinetic in regard to the experient (i.e. if this influence of "power" is the result of influencing the consciousness of the entity or some type of physical energy such as electric or luminous energy.

D. **Psychopompic Activity.** (Output) This specifier applies when an experient reports the ability to not only assist spirits, which have or have not crossed-over, resolve unfinished issues with the living, but also assist spirits that have not crossed-over on their journey to the afterlife.

E. **Compound.** Several modalities are involved, in which case each mode involved should be noted.

Associated Mental Health Findings

Mental health disorders somewhat common in experients of clairvoyant interaction include: Attention Deficit/ Hyperactivity Disorder, Depressive Disorder, Dissociative Disorder, Generalized Anxiety Disorder, Obsessive Compulsive Disorder, Panic Disorder with or without Agoraphobia, and Sleep Disorder (Kelly, 2011b).

Associated Medical Condition Findings

Physical medical conditions somewhat common in experients of clairvoyant interaction can include: Allergies/Sensitivities, Autoimmune Disorders, Chronic Fatigue Syndrome, Diabetes, Digestive Disorders, Hypertension, Hyperthyroidism, Hypoglycemia, and Epilepsy (Kelly 2011b).

Differential Classification

A variety of extrasensory and psychokinetic phenomena can present with somewhat similar phenomenology. These include:

o **Clairvoyant Cognition** or **Mental Mediumship**. Applied only when there is evidence to support that the information conveyed through the experient did not involve the occupation of another entity's consciousness or subjection to controlled behavior, but rather the information was only perceived by the experient.

o **Obsession.** Applied when reported features are similar to that of physical mediumship, but the effects are typically identified as the result of a single entity and are fairly continuous either in proximity to the experient or in proximity to the experient's home (i.e. recurring often over the course of weeks, months, or years). Here the physical manifestations are perceptible to other individuals and the experient may be the cause of an entity's increased ability to interact with its immediate environment. In other words, the experient is <u>haunted</u> by an entity.

o **Psychokinesis**. Applied when there is no evidence to support an "intelligent" entity was present during the physical influence of the environment (e.g. noises, rapping or knocking sounds, movement of objects, presence of heat or light without a known source, etc.). Rather, evidence supports that the experient may have been the solitary cause of the physical manifestations.

o **Apparition Activity**. Applied when a "ghostly image" is seen by an experient or the experient and other individual's that are the image of a deceased person or in some cases a living person in crisis (sometimes referred to as astral projection). However, this type of activity may also include other sensory-modalities besides the visual mode and can include more than the appearance of people or animals (living or deceased), but also inanimate objects. In the case of seeing a "ghostly image" of a discarnate entity, physical mediumship may still apply. This is of course unless the same apparition has been seen by other living individuals over time (e.g. an anniversary ghost; a ghost that appears the same time every year).

o **Poltergeist Activity**. Applied when experiences similar to physical mediumship are present, but effects appear to be limited to a certain location (residence, building, graveyard, etc.) and may involve the movement of objects (e.g. the throwing of objects), the displacement (e.g. the disappearance or appearance of objects, sometimes referred to as apportation), noises, voices, the sensation and/or physical evidence of biting, scratching, pushing, or pinching, the presence of unexplainable stains (e.g. blood), unexplained appearance of religious words, symbols, sigils, or images, or electronic phenomena/disturbances (e.g. lights flickering, power outages, etc.). However, if some of the aforementioned features are present and appear to be focused on the experient (e.g. the experient moved to a new residence and the entity followed) this could be an occasion of poltergeist obsession.

Criteria for Clairvoyant Interactive Experiences

A. **Characteristic phenomenology:** all of the following are required criteria for clairvoyant interactive experiences including criteria for clairvoyance in general.

> **(1)** Information is conveyed through the experient via some mode of entity interaction/occupation.
>
> **(2)** Information is conveyed in first person narrative (i.e. the entity is talking through the experient and refers to them elf (the entity) as "I").
>
> **(3)** Subconscious need for information conveyance present at the time of the experience.

2.3 Clairvoyant Simulation (CS)

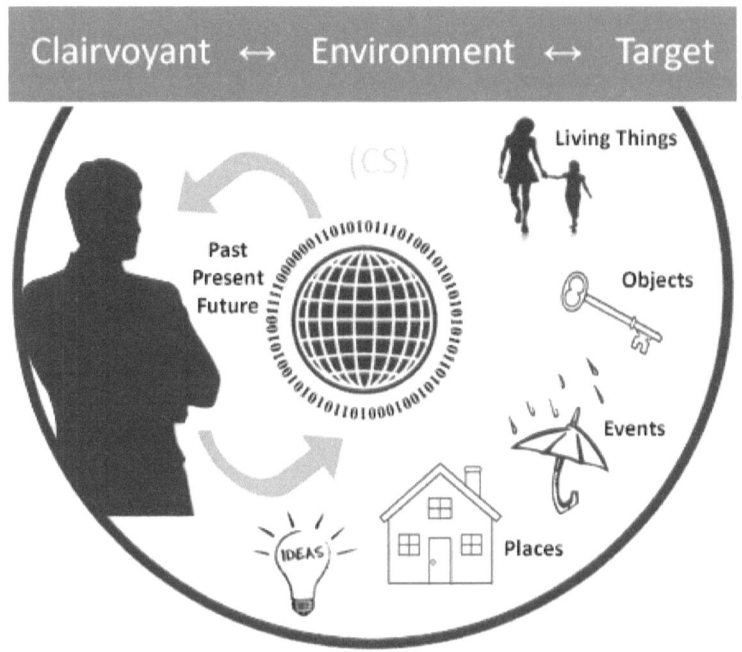

Clairvoyant ↔ Environment ↔ Target

Phenomenological Features

The essential feature of the (CS) type of clairvoyance is *a case in which an experient's mental or physical state appears to produce an accommodating effect in Nature, or Nature produces an accommodating effect within itself or the experient to satisfy the needs of the experient.* Clairvoyant simulation appears to involve the casual effective influence of an experient's mind on Nature, or Nature on the experient's mind or situation, without the intervention of the five senses. In other words, an experient's need produces an accommodating synchronistic (i.e. synchronicity) effect within Nature, or an experient's need for the production of an accommodating synchronistic effect occurs via Nature within or otherwise in regard to the experient (e.g. an idea or an "aha! factor) or within Nature. Through this type of clairvoyance, the experient does not "know," the information received is foreign (i.e. not of their own volition), nor is aware that their need is the source of their serendipity or luck (i.e. clairvoyant simulators typically believe they are simply "lucky," rather than "intuitive or psychic" due to the subtly of associated effects and the highly logical disposition typical of this type of experient).

The *input feature* of clairvoyant simulation allows Nature to "share" information with the experient that is accommodating to their current intentions or needs (e.g. acquiring information pertaining to *how*, or, *if* things work). Information acquired is typically additive to information already known to the experient, and as aforementioned, is typically acquired unbeknownst to the experient (e.g. ideas appear to be the product of the experients own deductions and conclusions, not "synthetic"). In other words, the information is typically presented as a self-concluded "aha!" or "eureka!" more so than as obtained from Nature. Therefore, this sudden realization is typically rarely construed as intuitive by the experient. As aforementioned, this type of clairvoyance appears to be more non-invasive then clairvoyant cognition or interaction, as the experient is typically unaware that, or does not "know" that, the accommodating information is "not their own." The skilled experient would however be able to identify that the simulated accommodating input information originated from Nature if the experient is made aware of his/her ability to share information with Nature, and is consistently "on the lookout" for synchronistic events (e.g. needed ideas that can be identified as void of a consecutive chain of thoughts or memories). This awareness is more likely of those with both clairvoyant simulative experiences and a more invasive type of extrasensory experiences (e.g. CC or TS) as associated sensations can be more easily identified in these experiences, and applied to detect more subtle CS sensations.

The *output feature* of clairvoyant simulation allows the experient to negotiate probability shifting with Nature, which is accommodating to their current intentions/needs (i.e. are more focused on shifting probability than information acquisition). Probability shifting is presumed as a psychical influence on Nature's stored probabilistic information (i.e. altering the probabilistic information pertaining to an object or event). Probabilistic information that appears to be shifted can include historical, real-time, or future probabilities (i.e. probabilistic information pertaining to past, current, or potential/future events). In regard to all of these types of influence, quantum information is neither created, copied, hidden, nor destroyed, but rather appears to be negotiated (e.g. an accommodating existing future potential is selected rather than various existing non-accommodating potentials).

Regardless of the type, probability is what appears to be shifted. However, in regard to different *temporal features* (i.e. past, present, or future), the "change" must have been, or currently is, a probability (i.e. however low the probability of the event, the event is nevertheless permitted by the Gaussian probability distributions of physical possibilities). Through clairvoyant simulation, the experient does not "know" what the accommodating effect will be per se, nor does Nature appear obligated to accommodate. Because of the very physical nature of clairvoyant simulation (e.g. accommodating output information simulation), questions have been raised as to if clairvoyant simulation should be considered a complex form of psychokinesis. Of course, ESP in general has also raised this same question (Kelly, 2011b). The differentiation between psychokinesis and clairvoyant simulation is addressed later in this section.

Intention Specifiers

The first set of specifiers is for identifying whether the experience was intentional or unintentional.

A. **Spontaneous.** This specifier applies when an accommodating effect occurs void of the experient's conscious choice, decision, or intention.

B. **Intentional.** This specifier applies when an accommodating effect occurs and the experient consciously selects or specifies the need for an accommodating effect, and is open to the occurrence of an accommodating effect.

The second set of specifiers is for identifying the subconscious or conscious need or goal that is assumed to be the catalyst for initiating clairvoyant simulative processes.

A. **Adaptive.** This specifier applies when information is shared or an accommodating effect is initiated to assist the experient in understanding and adapting, is initiated so Nature may adapt to the experients needs, or is initiated so other individuals within the experients situation may adapt to the experient's needs. The most common need is to provide social or emotional comfort and/or a sense of security. Here the experient typically has some level of emotional investment in the situation he/she requires information or an accommodating effect pertaining to (e.g. if the experient finds him/herself in a situation in which he/she does not feel comfortable or safe, Nature can share accommodating information that can alleviate anxiety, i.e. assist in conformity or change).

B. **Directive.** This specifier applies when information is shared or an accommodating effect is initiated that encourages activity that will lead to "good idea" or "good luck" or attract a helpful person or situation into the experient's life seen as luck or a meaningful coincidence. In this case, the experient and Nature are attempting to achieve a goal *with* each other. Therefore, Nature's "decision" to accommodate the experient's needs is assumed weighed against the situation and the needs of others within the situation. While Nature appears to be "friendly," it also appears to consider objectively (i.e. seems void of an individual bias). Here, Nature can, for example, share accommodating information that can lead to the alleviation of anxiety, i.e. impose direction).

The third set of specifiers is for identifying the direction of the clairvoyant simulative experience.

A. **Input**. This specifier applies when Nature shares information with the experient that is accommodating to their current intentions or needs (e.g. acquiring information pertaining to *how*, or, *if* things work). Information acquired is typically additive to information already known to the experient and is typically presented as a self-concluded "aha!" or "eureka!"

B. **Output**. This specifier applies when the experient reports probability shifting, whereby resulting in an effect accommodating to their current intentions/needs (i.e. are more focused on shifting probability than information acquisition).

The fourth set of intention specifiers is for identifying what types of synchronistic features where involved in the clairvoyant simulative experience. Typically one or more applies; therefore all that apply should be noted.

A. **Individual.** This specifier applies when one or more synchronistic features involve other individuals besides the experient (i.e. friend, family member, stranger, coworker, animal, insect, plant, etc.).

B. **Place.** This specifier applies when one or more synchronistic features are a physical location (e.g. park, school, home, restaurant, in the car, etc.).

C. **Object.** This specifier applies when one or more synchronistic features are inanimate physical objects (e.g. a money, number, book, computer, etc.).

D. **Idea.** This specifier applies when one or more synchronistic features are ideas, concepts, ideals, states, emotions, or attributes (e.g. thought, knowledge, question, anger, compassion, charity, etc.). Ideas can be further classified by identifying the nature of the idea, whether it be: coordinating (i.e. to systematically organize and arrange thoughts and ideas), constructing (i.e. to conceptualize; to construct hypotheses and theories based on known information), or investigating (i.e. to logically and systematically inquire as to the accuracy and validity of one's judgment, and as to the *how* or *if* systems work efficiently).

E. **Event.** This specifier applies when one or more synchronistic fea-
    tures are a situation (e.g. a business meeting, ceremony, expo,
    weather event, sports game, birthday, vacation, etc.).

The fifth set of specifiers is for identifying the temporal feature of the
clairvoyant simulative experience.

A. **Precognition.** This specifier applies when information acquisition
    pertains to the potential trajectory of a future state of an individu-
    al, object, or event. In other words, when probabilistic information
    pertaining to the future is perceived by an experient. This specifier
    includes presentiment and premonitions (i.e. where information
    perceived in regard to future events involves only emotional con-
    tent or pre-stimulus physical sensations that can be measured
    monitoring heart rate, brain waves, skin conductance, etc.).

B. **Probabilistic Shifting.** This specifier applies when probability shift-
    ing pertains to the potential trajectory of a future state of an indi-
    vidual, object, or event. Example, if existing probabilities are 1,2,3,
    and 4, and the most likely to occur is 1, but 2 is found to be more
    accommodating, then 2 will occur if simulation is successful.

C. **Contemporaneous.** This specifier applies when information acqui-
    sition pertains to the present/current state of a distant or unseen
    individual, object, or event (i.e. information pertaining to real-time
    states). Here, information acquisition is in real time, rather than
    occurring in a matter of hours, days, or months which is consistent
    with the precognition specifier, or in which has already occurred
    as of a matter of hours, days, or months which is consistent with
    the postcognition.

D. **Real-Time Shifting.** This specifier applies when shifting is in real-
    time. In other words, when the effect was immediate rather than
    delayed (e.g. for hours, days, months, etc.).

E. **Retro/Postcognition.** This specifier applies when information ac-
    quisition pertains to the past state of an individual, object, or
    event (i.e. information pertaining to historical states that may or
    may not be recent, but are not current).

F. **Historical Shifting.** This specifier applies when probability shifting pertains to the past state of an individual, object, or event (i.e. shifting pertaining to historical states that may or may not be recent, but are not current). Also known as retroactive psychokinesis (i.e. an instance of retroactive causation). Example, Event A only occurred because of historical shifting and Event A would not have occurred in this same manner if not for the following shift. In other words, the experient in the present time reaches back to alter the past. This is not to be confused with the concept of an experient in the future reaching back to alter our current time or our past, as this would require very different phenomenology, theories of time, and require a purely deterministic philosophical view.

Development and Course

Childhood onset may present itself through compound modalities including dream, intuitive impressions, auditory, visual, and other modes. Adolescent onset primarily presents itself most commonly through auditory, visual, other, and compound modes, with dream and intuitive impressions being less common. Adult onset primarily presents itself most commonly through dreams and intuitive impressions, with auditory, visual, and other modes being less common.

Course Specifiers

These specifiers are for identifying the characteristic course of clairvoyant simulative experiences over time.

A. **Single Episode.** This specifier applies when the experient reports a clairvoyant simulative episode without a prior history of episodes.

B. **Episodic.** This specifier applies when the experient reports clairvoyant simulative episodes of which seem to occur irregularly and of which the duration of the experience is momentary.

C. **Continuous.** This specifier applies when the experient reports clairvoyant simulative episodes of which seem to occur in a continual manner, or when episodes are so frequent it is difficult for the experient to determine where one episode ends and another begins (e.g. prolonged and closely spaced episodes).

Modality Specifiers

These specifiers are for identifying the characteristic mode(s) of clairvoyant simulative experiences. Here, some emotional investment in the situation on the experient's behalf is expected.

A. **Dream.** Refers to when one or more accommodating effects are experienced during sleep (e.g. Nature shares information with the experient during a dream).

B. **Intuitive Impressions/Emotional.** Refers to when one or more accommodating effects are experienced as emotional content (e.g. Nature shares information with the experient that shifts the experients perspective, whereby resulting in an emotional shift in the experient).

C. **Auditory.** Refers to when one or more accommodating effects are auditive (i.e. the experient hears an accommodating effect).

D. **Visual.** Refers to when one or more accommodating effects are visual (i.e. the experient sees an accommodating effect).

E. **Other.** Refers to when one or more accommodating effects do not seem to fit into any of the above categories.

F. **Compound.** Several modalities are involved, in which case each mode involved should be noted.

Associated Mental Health Findings

Mental health disorders somewhat common in experients of clairvoyant simulation include: Attention Deficit/ Hyperactivity Disorder; Bipolar Disorder, Depressive Disorder; Dissociative Disorder, Generalized Anxiety Disorder; Obsessive Compulsive Disorder; Panic Disorder with or without Agoraphobia, and in rare cases, Personality Disorder and Schizophrenia (Kelly, 2011b).

Associated Medical Condition Findings

Physical medical conditions somewhat common in experients of clairvoyant simulation can include: Absence Seizures, Diabetes, Electromagnetic Sensitivity, Hyperthyroidism, and Hypoglycemia (Kelly 2011b).

Differential Classification

A variety of extrasensory and psychokinetic phenomena can present with somewhat similar phenomenology. These include:

o **Clairvoyant Cognition**. Applied only when there is evidence to support that the information was "known" rather than shared, or was in third-person perspective or narrative rather than seemingly having no perspective. This also applies to reports that are the result of divinatory practices.

o **Clairvoyant Interaction**. Applied when a sharing-like process is reported, but motor control was compromised during the process suggesting automatism.

o **Telepathic Simulation.** Applied if another individual is assumed to have been a participant in the sharing process rather than the environment (i.e. evidence to support mind-to-mind communication).

o **Precognition**. Applied when information pertaining to future events is "known" rather than shared, is not considered additive to information previously known to the experient, or the event is not accommodating. However this does not apply when an event preceding another is reported as a meaningful coincidence, and in retrospect, could be seen as a "for telling" of an event to come.

o **Psychokinesis**. Applied when an output effect is not accommodating, involves the <u>direct</u> physical influence of an individual (e.g. healing), or involves the <u>direct</u> physical influence of an object (e.g. movement). Psychokinesis involves energy and physical influence such as heating, cooling, illuminating, movement, levitation, etc. with very small effect sizes, while clairvoyant simulation involves primarily an effect on probability information, which can result in small to seemingly very large effect sizes. Example, if an individual is about to be assaulted and the assailant falters due to a sudden electrical pain sensation the their body, this is more likely to be the cause of PK rather than CS (i.e. bioelectric). However, if an individual is about to experience a car accident and suddenly the most unlikely scenario plays out in which spares the experient of being in an accident, this is more likely to be the cause of CS rather than PK (i.e. shifting to the least likely probability). Example, if an individual is healed of a case of acute inflammation, this may be the result of PK (i.e. bio-energy healing/repair, where only the experient is involved in the healing process). However, if the individual is healed of a congenital disorder (i.e. a condition existing before birth, at birth, or that develops during the first month of regardless of causation), this may be the result of CS (i.e. biological/genetic information shifting, where the experient and Nature are the source of the healing process).

Criteria for Clairvoyant Simulative Experiences

A. **Characteristic phenomenology (Input):** all of the following are required criteria for clairvoyant simulative experiences with the input feature including criteria for clairvoyance in general.

   (1) Accommodating information is shared between the experient and the environment/Nature.

(2) Accommodating information is shared with seemingly no perspective other than the self (e.g. if visual: the image is shown from the experient's perspective, if auditory: information is heard in-mind in the experient's own voice and is heard in first-person narrative such as "I think," or "I feel."

(3) Subconscious need for accommodating information present at the time of the experience.

B. Characteristic phenomenology (Output): all of the following are required criteria for clairvoyant simulative experiences with the output feature including criteria for clairvoyance in general.

(1) An accommodating and meaningful coincidence, synchronistic event, or probability shift has occurred.

(2) Subconscious need for an accommodating meaningful coincidence, synchronistic event, or probability shift present at the time of the experience.

3. EMPATHY

Mind-to-Mind / -Environment Emotional Communication

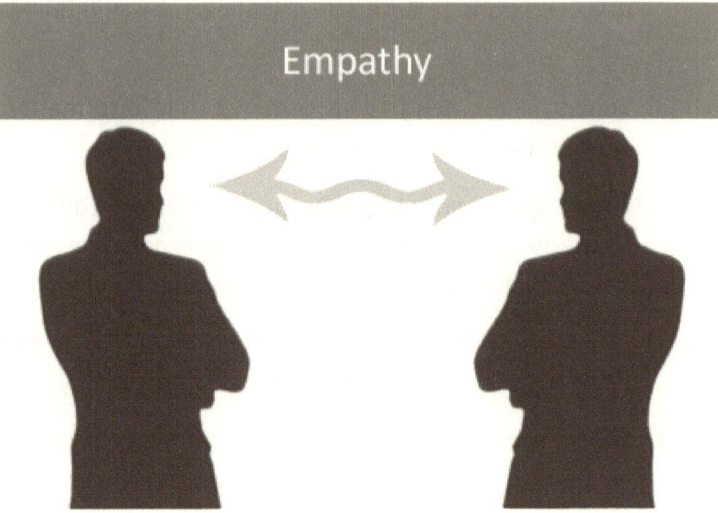

Phenomenological Features

The extrasensory experiences in this section include Generalized Empathy, Empathic Cognition, Empathic Interaction, and Empathic Simulation. These experiences have been grouped together to facilitate differential classification of experiences that include Empathy as a prominent aspect of the experience. The term *empathy* in the parapsychological context is defined as *the psychical influence of emotion via experient influence over the emotional basis of consciousness and the mental and physiological processes associated with a wide variety of emotional experiences*. Emotions are defined as an episode, which suggests the concept of a dynamic process, of interrelated, synchronized changes in the states of all or most of the correlated organismic subsystems (e.g. central, neuroendocrine, and somatic nervous systems) in response to the evaluation of an external or internal stimulus event as relevant to major concerns of the organism.

The function of emotion is speculated to include the evaluation of objects and events, system regulation, preparation and direction of action, communication of reaction and behavioral intention, and the monitoring of internal state and organism-environment interaction. Current thought leaders in regard to the psychology of emotions support a component process model of emotion involving cognitive, neurophysiological, motivational, motor expression, and subjective feeling components. In addition, empathy can involve the influence of affective phenomena such as moods. However, this appears to apply only when an element of telepathy or clairvoyance is involved, as affective phenomena such as moods involve more than just emotional content. It is an experient of empathic experience's natural endowment in which enables their influence over emotion whether the experience is the result of conscious or subconscious performance (Kelly, 2011c).

Subtypes

The following subtypes are phenomenological subgroups exclusive to empathy only.

3.1 (EC)   **Empathic Cognition** (see p. 98)
3.2 (EI)   **Empathic Interaction** (see. p. 107)
3.3 (ES)   **Empathic Simulation** (see p. 115)

Severity Specifiers

These specifiers should only be used when all criteria for the type of a subtype are currently met. In deciding whether reported experiences should be described as stable/functional, mild, moderate, or severe, the clinician should take into account the number and intensity of the experiences and any resulting impairment in occupational or social functioning.

A. **Stable/Functional.** Intentional experiences of which fit all criteria with few, if any, spontaneous experiences and of which result in no impairment in social or occupational functioning and may or may not increase normal functioning.

B. **Mild.** Few experiences of which fit all criteria and experiences result in no more than minor impairment in social or occupational functioning.

C. **Moderate.** Experiences and functional impairment between "mild" and "severe" are present.

D. **Severe.** Many experiences of which fit all criteria, either episodic or continuous, of which result in marked impairment in social or occupational functioning.

Associated Research and Laboratory Findings

No laboratory findings have been identified that are diagnostic of empathy in a parapsychological context. However, a variety of measures from neuroimaging, neuropsychological, and neurophysiological studies have shown differences between groups of individuals with empathy and appropriately matched control subjects. In a study by Radin & Schlitz (2005), investigating whether the emotions of one individual, measured with an electrogastrogram (EGG), respond to the emotions of a distant individual; EGG maximums were significantly larger when the distant individual was experiencing positive emotions, larger when experiencing negative emotions, and smaller when experiencing calm emotions as opposed to neutral emotions. When focusing on negative emotions, EGG maximums for sadness rather than anger was found to be significantly larger than neutral.

Specific Culture, Age, and Gender Features

Clinicians assessing beliefs and claims in socioeconomic or cultural situations that are dissimilar from their own must take cultural dissimilarities into account. Ideas that may appear to be questionable or even delusional in one culture or subculture (e.g. Buddhists, New Agers, Wiccan Practitioners, and those who engage in regular meditative practices) may be commonly believed in another. In some cultures, empathic impressions or intuitions with a spiritual or religious content

may be a normal part of spiritual or religious experience (e.g. the sensing of a loved one in distress, mothers intuition; suggestive of a empathic or telepathic connection between mother and child often seen
in a spiritual context). These varying beliefs may have subtle to blatant
differences in terminology and descriptions leaving the clinician with
the difficult task of properly categorizing experiences into parapsychological types and subtypes.

In regard to physical location, in a study conducted by Haraldsson
& Houtkooper (1991), individuals in multiple countries reported telepathy and clairvoyance experiences. While this study does not focus on
reports of purely emotional extrasensory experiences, its percentages
may be used in conjunction with typical distributions of purely intuitive extrasensory experiences. According to Irwin (2007), "intuitive
impressions may include some appreciation of the identity of the person whom, or the situation to which, the felt emotion relates." However, Irwin also states that "in occasional instances the information
element is minimal and the experience comprises little else than a
strong, unexpected, emotion." According to Rhine (1951), intuitive
impressions account for 26% of extrasensory experience, with 9% hallucinatory, 44% realistic dreams, and 21% unrealistic dreams. Taking
all of the above into consideration, similar to intuitive impressions,
which in some cases are defined synonymously with empathic experiences, purely empathic extrasensory experiences are reported more
often than telepathic or clairvoyant hallucinatory experiences, but less
often than intuitive impressions involving more information than emotional content and basic association.

Initial experiences (onset) of empathic experiences typically occur
within the first several years after birth and/or during puberty. Early
onset may involve many spontaneous experiences of which may or
may not affect the child psychologically, emotionally, or socially. Experiences in which have an early onset and continue throughout life
without extended pause (e.g. 1 year or more without an experience)
typically remain stable/functional in the long term. In some generalized ESP experiments, children tend to score higher than adolescents
and adults. However, many similar studies have been unsuccessful in

in demonstrating age dependent differences in scoring (Palmer, 1978). According to Blackmore (1980), "Although many studies show high scoring in children there is little systematic evidence of a relationship between ESP and age and there are many contradictory findings."

The onset of empathic experiences during puberty, most common between the ages of 13-16, is typically induced to compensate for an inability to effectively communicate their wants, needs, thoughts, and/or emotions verbally to other individuals, or lack individuals in their life that could properly meet their needs. Experients may feel they have had a recent decline in quality of life, academic perfor- mance, and/or social relationships. During this time experiences are typically spontaneous, and can range from mild to severe depending on the severity of needs the experient feels they are unable to com- municate and obtain. Experiences in which are moderate to severe have a high probability of continuing in severity unless the want or needs of the adolescent are met. Onset during this age rage may also be induced by another individual unwilling to meet the experients wants or needs, or the adolescent's general environment may be un- accommodating in some manner.

In other words, the experient is communicating effectively, but (1) the recipient does not understand effectively (e.g. a parent that lacks emotional intelligence i.e. cannot effectively understand and respond- ed to a child in emotional distress, or where the parent is the source of emotional distress and does not understand the part they play in the child's emotional instability), or (2) simply refuses to meet the experi- ents needs (e.g. a bully at school, or an emotionally abusive or neglect- ful parental figure), or (3) the experient is able to communicate effec- tively, but the resources required are not being made available (e.g. emotional support from family and friends, education in emotion- focused coping/regulation skills, etc.)

Adult onset may occur at any age and is typically precipitated by and inability to verbally communicate wants, needs, thoughts, or emo- tions, physically acquirer wants or needs, or deepen an emotional connection with another individual (e.g. an individual in the experients life or a type of individual the experient wants in their life). Adult on-set is typically stable/functional to mild unless precipitated by experiences that amount to trauma, illness, or any other type of sudden un-complimentary experience, acute or chronic, that results in a major disturbance in the experient's life. In the case of the latter, moderate to severe experiences are typically common. Spontaneous experiences are common regardless of the severity. However, stable/functional to mild experiences are more likely to be the product of intention, while moderate to severe experiences are mainly spontaneous.

Gender differences have been the focus of some studies. Overall, there appears to be no clear trend for differential scoring between males and females (Palmer, 1978), and if gender differences are found, they tend to be slight with women reporting empathic experiences more than men, men reporting empathic experiences more than women, or no difference between gender reporting was found. In addition, there has been evidence supporting that mixed-gender pairings (empathist and subject/participant) are more successful than same-gender pairings (Dalton & Utts, 1995).

Familial Patterns

Occasionally one biological parent or grandparent of an experient of empathic experiences reports a history of empathic-like experiences. Familial patterns most common are empathic experiences between mother and child, spouses/lovers, identical twins, and occasionally between fraternal twins, siblings, and meditation partners. In regard to marital status in Europe and the U.S., relatively fewer single and married individuals report telepathy, clairvoyance, and contact with the dead then the "combined broken relationship group" (i.e. living as married, separated, or divorced) (Haraldsson & Houtkooper, 1991). As aforementioned, while this study does not focus on reports of purely emotional extrasensory experiences, its percentages may be used in conjunction with typical distributions of purely intuitive extrasensory experiences.

Associated Terminology

Emotional content. Experients of empathy in which sense only emotional content may use the following terminology: clair-empathic • clair-empathy • empath • empathic • empathist • empathy • gut feeling • intuition • intuitionism • intuitive • intuitvism • keen intuition • presentiment • mothers intuition • sensitive • tele-empathy • tele-empathic.

Additional terminology. Used in a context involving the experient and at least one other individual, or social group, or the environment, including both telepathy and telepathy-like terminology: anomalous emotion communication • emotion reading • emotion reception • emotion transference • emotion or mood transmission • emotion withdrawal • emotional compulsion • emotional control • emotional influence • emotional suggestions • insight • psychic communication • psychic knowledge • six sense • twin empathy.

Criteria for Empathic Experiences

A. **Characteristic phenomenology:** all of the following are required criteria for empathic experiences.

   (1) Mind-to-Mind, or mind-to-environment, emotional communication.

   (2) Involves one or more individuals, or involves one or more environments and indirect emotional information pertaining to one or more groups of individuals.

   (3) The subject or participant is a living organism (e.g. human, animal), or the target group is comprised of living organisms (e.g. human, animal) and the information obtained about them is in reference to the emotional state of the target group (e.g. emotions towards community or national health, politics, current events, etc.).

B. **Social/occupational need:** A subconscious need has been identi-
    fied as the catalyst for the initiation of empathic processes (i.e. (1)
    identified an inability to communicate wants, needs, or emotions
    to an individual(s) in an interpersonal, academic, or occupational
    context, (2) inability to acquire physical necessities, (4) need for
    emotional information not readily accessible though natural
    means, etc.).

C. **Validation:** The experience has been validated by an individual
    other than the experient (e.g. the subject(s) confirmed the accura-
    cy of the emotional information received by the experient, or a
    reputable news source confirmed the accuracy of the information
    received by the experient (e.g. confirms a tragic accident affecting
    a particular group of individuals after the experient reported an in-
    tense sadness originating from that particular group, or at least
    identified the sadness as not originating from the empathist), and
    the clinician determines the experience was more than a coinci-
    dence/chance occurrence based on the quality of the emotional
    information received and reported, and all other possible explana-
    tions for obtaining the emotional information is excluded. If vali-
    dation does not apply, yet empathic processes are still plausible,
    the experience should be assigned as "Possible Empathy" (PE).

D. **Clairvoyance and Telepathy Exclusion:** Clairvoyance and telepathy
    have been ruled out because no more than emotional content has
    been identified as the basis of the experience(s) other than a
    sense of where the emotional content originated.

Empathic Subtypes

The subtypes of Empathy are defined by the predominant phenomenology of reports. The determination of a particular subtype is based on the clinical picture that occasioned the most recent experiences, and may therefore change over time. Not infrequently, the description of experiences may include phenomena that are characteristic of more than one subtype. The choice among subtypes depends on the following algorithm: Empathic Cognition (EC) is assigned whenever emotional information is clairvoyantly acquired by the empathist originating from an environment/situation pertaining to a group of individuals; Empathic Interaction (EI) is assigned whenever emotional information is empathically acquired by a subject originating from the empathist; Empathic Simulation (ES) is assigned whenever information is shared between the empathist and a participant; If two or three subtypes are assigned, all should be listed; Generalized Empathy (GE) is assigned when all subtypes appear to apply (optional), or a clear choice is unable to be made, but appears to only suggest an empathic experience. In addition, when a clear choice cannot be made, the clinician should consider a dimensional approach to classifying the experiences.

3.1 Empathic Cognition (EC)

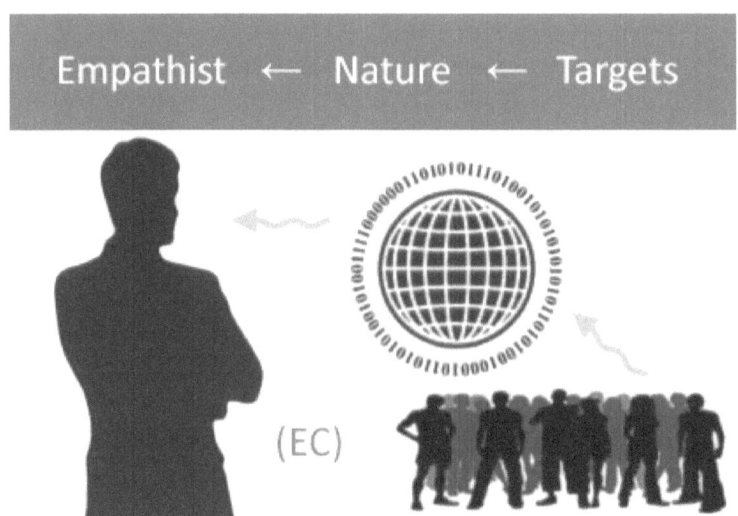

Empathist ← Nature ← Targets

(EC)

Phenomenological Features

The essential feature of the (EC) type of empathy is the *phenomeno-logically indirect knowledge of the collective emotional experience of a large group or population via Nature*. In cases of Empathic Cognition, or Empathic Clairvoyant Cognition, an experient is retrieving information from the environment or "Nature" (i.e. the experient is able to "pick up on" a current collective emotional experience recorded in Nature). This is to say, that even if the emotional experience is current, it is assumed that Nature "knows" about the experience, and therefore has a "record" of the experience. The informational system from which the collective emotional experience originates does appear to play an intentional part in the emotional information teleportation processes. In other words, the process is not assumed entirely evoked by the empathist, whereby possibly suggesting Nature's ever-present contribution to the survival (well-being) of the empathist.

Again, in regard to empathic clairvoyant cognition, the experient is an individual with clair-empathic ability capable of evoking clair-empathic processes, or allowing the occasion for Nature to evoke such processes when needed by the empathist. Here, the empathist will become aware of Nature-based information (e.g. collective emotional experiences), but should be able to clearly identify that the information did not originate in their own mind. Here the emotional information is received and perceived by the empathist, but the information did not develop from a chain of prior emotions belonging to the empathist. Instead, the emotional experience appears to "pop up," but is immediately associated with a specific group or populous, or simply identified as not originating from the empathist.

The type of experient of empathic clairvoyant cognition described above could be defined as an experient of *spontaneous* empathic clairvoyant cognition, in that the emotional experience appears to "pop up" rather than being intentionally requested. Empathists of *intentional* empathic clairvoyant cognition are experients whom select or specify a particular group from which they wish to extract collective emotional information regarding. However, this type of empathist can also generalize their search from a "which (e.g. group)," to "what (e.g. emotion or event)" -based search depending on the type of collective emotional information the empathist requires.

In the case of the latter, it is a search for collective emotional events that is initiated (i.e. when a large group of typically thousands or millions of individual emotional experiences, or consciousness, becomes coherent and synchronized). In other words, when a unified field of consciousness occurs, often void of information pertaining to what the event is or where unless there is a telepathic component, an empathic clairvoyant cognitive can acquire emotional information pertaining to the event when such an event takes place. This search type is sometimes interpreted as a form of precognition, but rather appears to be contemporaneous (i.e. contemporaneous ESP) since the collective emotions are only "known" to the empathist at the time of the event.

In regard to the scientific research of emotions, an *emotional climate* is defined as the emotional relationships between members of a society or nation. Emotional climates are assumed to contribute to maintaining the political unity or cultural identity of the members of a society or nation. Emotional climates appear to be emergent process- es of which have a social function and are formed by an "aggregate of the four 'basic' emotions" (i.e. fear, anger, sadness, joy with an ac- companiment of "specific ways of social interaction" and "specific predisposition towards action"). While an emotional climate appears to be based on more than just emotions (e.g. beliefs, social represen- tations, etc.), it is assumed to be what empathic clairvoyant cognitive processes are additive to. These collective phenomena are not be- lieved to be simply an aggregation of individual emotions or all emo- tional relations, but rather collective phenomena based on the pre- dominance of particular emotions, which are what is assumed to be acquired by the empathist (Kelly, 2012).

Intention Specifiers

The first set of specifiers is for identifying whether the experience was intentional or unintentional.

A. **Spontaneous.** This specifier applies when emotional information appears to "pop into mind" rather than being intentionally re- quested by the percipient.

B. **Intentional.** This specifier applies when a percipient selects or specifies the source (i.e. group or populous) of emotional infor- mation required.

The second set of specifiers is for identifying the subconscious or con- scious need or goal that is assumed to be the catalyst for initiating empathic cognitive processes.

A. **Adaptive.** This specifier applies when emotional information ac- quisition is initiated to assist the percipient in understanding and/or adapting to a group or populous of individuals to which they typically have some level of emotional investment.

B. **Decisive.** This specifier applies when emotional information acquisition is initiated to assist the percipient in coming to a decision involving a group or populous of individuals to which they typically have some level of emotional investment.

Course Specifiers

These specifiers are for identifying the characteristic course of empathic clairvoyant cognitive experiences over time.

A. **Single Episode.** This specifier applies when the percipient experiences empathically acquired emotional information without a prior history of episodes.

B. **Episodic.** This specifier applies when the percipient experiences empathically acquired emotional information of which seems to occur irregularly and of which the duration of the experience is very momentary. While the percipient may appear distracted during an empathic clairvoyant cognitive episode, the experient should still be fully aware of their surroundings.

C. **Continuous.** This specifier applies when the percipient experiences empathically acquired emotional information of which seems to occur in a continual manner, or when episodes are so frequent it is difficult for the percipient to determine where one episode ends and another begins (e.g. prolonged and closely spaced episodes).

Modality Specifiers

This specifier is for identifying the characteristic mode(s) and submode(s) of an empathic clairvoyant cognitive experience. In any case, some emotional investment in the group or populous of individuals, or the situation in which the group or populous of individuals reside, on the experients behalf is expected.

A. **Dream.** Refers to empathic information acquisition during sleep.

B. **Intuitive Impressions/Emotional.** Refers to non-hallucinatory sensations of which can be described as empathically received emotional content during normal or altered (e.g. trance) states of conscious awareness.

    a. **Achievement Emotions**. Refers to the class of utilitarian emotions including <u>pride</u> (i.e. associated with an enhancement of ego-identity and self-esteem), <u>elation</u> (i.e. provides an individual with the feeling of living fully), <u>joy</u> (i.e. elicits confidence, comfort, and boosts self-esteem), and <u>satisfaction</u> (i.e. is contributing to a feeling of fulfillment and wellbeing).

    b. **Approach Emotions.** Refers to the class of utilitarian emotions including <u>relief</u> (i.e. results subsequent to a negative emotion when an event has taken a turn for the betterment of the individual or group), <u>hope</u> (i.e. contains some level of uncertainty because it is future orientated, but plays a vital role in adaptation as a means to pursue ones goals), <u>interest</u> (i.e. the emotion elicited when one experiences a feeling of engagement, fascination, and curiosity), and <u>surprise</u> (i.e. accompanied by uncertainty, which keeps one on their toes and stimulated as a means to cope with and adjust to new and unexpected actions and events).

    c. **Resignation Emotions.** Refers to the class of utilitarian emotions including <u>sadness</u> (i.e. evoked when one loses something in life such as a loved one, employment, or social standing, and is typically correlated with resignation and failure), <u>fear</u> (i.e. activates a sense of threat, or uncontrollability, alongside a need to preserve integrity), <u>shame</u> (i.e. the emotion experienced when a negative appraisal of the all-inclusive self is concerned, when an individual experiences humiliation or feels as though others find the individual insignificant or worthless, and can result in a momentary inability to think logically and efficiently), and guilt (i.e. associated with negative self-appraisal, but unlike shame, it is related to specific actions and behaviors).

d. **Antagonistic Emotions.** Refers to the class of utilitarian emotions including envy (i.e. the emotion evoked when an individual desires what another individual possesses and feels a sense of inferiority in comparison to the other individual), disgust (i.e. the emotion experienced when an individual is confronting something considered repulsive or abhorrent), contempt (i.e. the emotion experienced when an individual feels superior and dominant, but typically does not engage in aggressive behaviors such as assault), and anger (i.e. the emotion experienced when one feels a sense of wrongdoing, on their part or by other individuals, which is considered offensive and is accompanied by a sense of injustice, unfairness, or inequality).

e. **Aesthetic Emotions.** Refers to the class of emotions including those characteristic of an absence or a less pronounced function for immediate adaptation to a situation of which requires goal relevance evaluation and coping potential. In other words, the aesthetic experience of a work of art or a piece of music is not formed through the appraisal of whether the work meets physical needs, assists in furthering current goals or projects, or is in accordance with personal social values. Instead, aesthetic emotions are the product of an appreciation for the intrinsic qualities of naturalistic beauty, or the qualities of a work of art or artistic performance. A few examples of aesthetic emotions include being moved, in awe, full of wonder, admiration, bliss, ecstasy, fascination, harmony, rapture, and solemnity.

    f.   **Compound.** Several submodalities or emotions are involved, in which case each submode and identifiable emotion involved should be noted.

Sub-Subtype Specifiers

These specifiers are for identifying additional sub-subtypes specific to empathic cognition. Some emotional investment in the group or populous of individuals on the experients behalf is expected.

A. **People-Orientated.** This specifier applies when emotional information is acquired from Nature, the emotional information is predominately in reference to a biological organism of interest to the percipient (e.g. are within their own environment). This environment *may* include a single individual, though single emotional experiences are not the primary focus of these types of empathists, but typically includes those in *their* neighborhood, workplace, community, etc. In other words, this subtype is typically only concerned with obtaining emotional information that is exclusively relative to the percipient as a means to empathize with others in their surround.

B. **Nature-Orientated.** This specifier applies when percipients have a primary focus on acquiring information pertaining to large scale emotional events (i.e. major events and mass emotions) that pertain to biological organisms that may or may not be on an entirely different side of the globe (i.e. not within their own more immediate environment). Therefore, the primary distinguishability between these two sub-subtypes is an interest in a familiar group's collective emotions (i.e. which), and an interest in large-scale, national or global, emotional events evoked by, for example, natural or unnatural disasters (i.e. "which" rather than "who").

Associated Mental Health Findings

Mental health disorders somewhat common in experients of empathic cognition include: Depressive Disorder; Generalized Anxiety Disorder;

Obsessive Compulsive Disorder; and Panic Disorder with or without Agoraphobia (Kelly, 2012).

Associated Medical Condition Findings

Physical medical conditions somewhat common in experients of empathic cognition can include: Diabetes, Hypertension, Lupus, Lymphatic System Disorders, and rarely Meningitis (Kelly 2011b).

Differential Classification

A wide variety of extrasensory phenomena can present with similar phenomenology. These include:

o **Clairvoyance** and **Telepathy**. Applied when more than emotional content is perceived such as images, sounds, and other sensations. If the experient reports some purely empathic experiences and some experiences involving more than emotional content, the experient should be considered clairvoyant or telepathic with occasional intuitive impressions and/or hallucinations.

o **Empathic Simulation.** Applied when the emotional state is "shared" rather than "known." Empathic simulation is often misinterpreted as empathic cognition as both can involve large groups. Differentiating between indirect (i.e. group » Nature » experient) and direct (i.e. subjects » experient) information can be achieved by identifying whether the emotional state was known (i.e. the experient was cognizant or aware of the groups emotional state and could identify that the emotional experience was "not their own"), or shared (i.e. the experient experienced the emotional state of the group).

o **Precognition**. Applied if the emotions felt were prior to the actual event in which stimulates the emotional response. In this case, the experient should be considered clairvoyant rather than empathic, as empathic cognition at this time appears to be exclusively con-temporaneous.

Criteria for Empathic Cognitive Experiences

A. **Characteristic phenomenology:** all of the following are required criteria for empathic cognitive experiences including criteria for empathy in general.

   (1) Emotional information is received by the percipient through mind-to-environment communication.

   (2) Involves one or more environments and indirect emotion-al information pertaining to one or more groups of indi-viduals.

   (3) The source of the information is not a single living organ-ism (e.g. human, animal), but rather the information ob-tained is in reference to the emotional state of a group of living organisms of which the emotions of a single individ-ual may be identifiable (e.g. emotions towards communi-ty, national health, politics, current events, etc.)

   (4) Subconscious need for emotional information acquisition present at the time of the experience.

3.2 Empathic Interaction (EI)

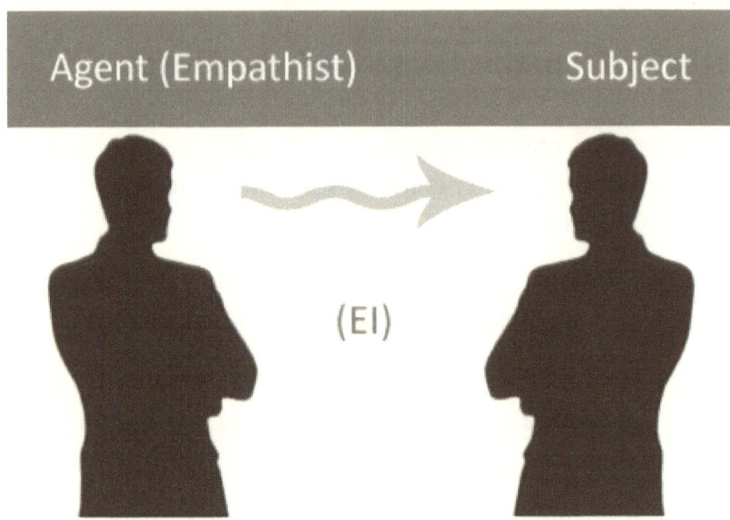

Phenomenological Features

The essential feature of the (EI) type of empathy is *the causal influence of one mind on another's emotional state without the intervention of the five senses*. Through empathic interaction, the empathic impressionist influences the emotional state of another as a means to instruct subjects to feel a particular way that is accommodating to the empathist. However, it appears that there may be an associated hypnotic element in regard to empathic interaction. *Hypnogenic empathic interaction* is a form of empathic interaction that is assumed involved in causing a mild hypnotic state in subjects, possibly involving impressed emotions related to relief, which results in the behaviors of relaxation or decompression, via an empathists command.

This relaxed state allows for a more dominant approach by the empathist in influencing the emotions of others, and allows subjects to be more susceptible to impression. Hypnogenic empathy appears to not only evoke strong emotions in subjects, but also typically results in prompting a behavior in subjects. Therefore, hypnotic empathic interaction appears to be the strongest form of empathy and the most dangerous, raising an assortment of moral and ethical questions as to how such an ability should be utilized in practical applications. Further studies suggest that initial empathic "impressions" (i.e. commands/evoked emotions) do not always fade away with time, but rather occasionally result in the same strength of emotion anytime a subject is confronted by the empathist, or in some cases are merely reminded of the empathist (Kelly, 2012).

Intention Specifiers

The first set of specifiers is for identifying whether the experience was intentional or unintentional.

A. **Spontaneous.** This specifier applies when the empathist impresses the subject void of the empathist's conscious choice, decision, or intention.

B. **Intentional.** This specifier applies when an empathist selects or specifies a subject to be impressed with emotional information or is coerced into action or the expression of an emotion. However, this specifier also applies when an empathist generalizes their interaction from "who" to "what" the empathist needs from the subject to result in a modification of a situation involving the subject. In the case of "what," one or more subjects may be impressed.

The second set of specifiers is for identifying the subconscious or conscious need or goal that is assumed to be the catalyst for initiating empathic interactive processes.

A. **Adaptive.** This specifier applies when emotional information impression is initiated to assist the subject in emotionally understanding and emotionally adapting to the empathist's needs or goals. Here the empathist typically has some level of emotional investment in the subject or the situation in which the subject occupies.

B. **Directive.** This specifier applies when emotional information impression is initiated to assist the subject in complying with a suggestion or command to feel a specific emotion. This is done in expectation that the subject will be provoked into acting or behaving in a specific or generalized manner as a direct result of that emotional state. Behaviors are affectional (i.e. actions which are taken due to one's emotions, to express personal feelings). Here the empathist typically has some level of emotional investment in the subject or the situation in which the subject occupies.

The third set of specifiers is for identifying the type of impression experienced by the subject.

A. **Suggestive.** This specifier applies when the emotional information impressed can be identified as a proposal of which may be accepted or rejected per the subject's prerogative.

B. **Compulsive.** This specifier applies when the emotional information impressed can be identified as either coercive towards irrational feelings, or coercive towards rational feelings, but the subject feels they are behaving against their will or by force.

Course Specifiers

These specifiers are for identifying the characteristic course of empathic interactive experiences over time.

A. **Single Episode.** This specifier applies when the empathist impresses an emotion onto a subject and the empathist and subject report no prior history of episodes. This specifier also applies when the empathist suggestively or compulsively evokes an emotion in the subject, which provokes the subject into action or towards behaviors that the subject reports are not typical (i.e. the subject has not responded in such a way in similar circumstances in the past), but the empathist and subject report no prior history of episodes.

B. **Episodic.** This specifier applies when the empathist impresses an emotion onto a subject, which seems to occur irregularly and of which the duration of the experience is very momentary. An episodic emotional impression may involve a quick burst of emotion with the duration of the experience lasting only a maximum of a couple of seconds. This specifier also applies when the empathist suggestively or compulsively evokes an emotion in the subject, which provokes the subject into action or towards behaviors that are not typical of the subject in the past, but now the subject irregularly acts or behaves in such a manner.

C. **Continuous.** This specifier applies when the empathist impresses emotional impressions onto a subject of which seem to occur in a continual manner, or when episodes are so frequent it is difficult for the empathist or subject to determine where one episode ends and another begins (e.g. prolonged and closely spaced episodes).

Modality Specifiers

This specifier is for identifying the characteristic mode(s) and submode(s) of an empathic experience. In any case, some emotional investment in the subject, or the situation in which the subject resides, on the empathist's behalf is expected.

A. **Dream.** Refers to empathic impression during sleep where the subject is impressed with emotional information during the dream state and/or engages in an action or behavior once awake due to the impressed emotional content of the dream.

B. **Intuitive Impressions/Emotional.** Refers empathic impression during normal or altered (e.g. trance) states of conscious awareness, of which can be described as emotional content impressed onto a subject that may result in an action or behavior.

   a. **Achievement Emotions.** Refers to the class of utilitarian emotions including pride (i.e. associated with an enhancement of ego-identity and self-esteem), elation (i.e.

provides an individual with the feeling of living fully), joy (i.e. elicits confidence, comfort, and boosts self-esteem), and satisfaction (i.e. is contributing to a feeling of fulfillment and wellbeing).

b. **Approach Emotions.** Refers to the class of utilitarian emotions including relief (i.e. results subsequent to a negative emotion when an event has taken a turn for the betterment of the individual or group), hope (i.e. contains some level of uncertainty because it is future orientated, but plays a vital role in adaptation as a means to pursue ones goals), interest (i.e. the emotion elicited when one experiences a feeling of engagement, fascination, and curiosity), and surprise (i.e. accompanied by uncertainty, which keeps one on their toes and stimulated as a means to cope with and adjust to new and unexpected actions and events).

c. **Resignation Emotions.** Refers to the class of utilitarian emotions including sadness (i.e. evoked when one loses something in life such as a loved one, employment, or social standing, and is typically correlated with resignation and failure), fear (i.e. activates a sense of threat, or uncontrollability, alongside a need to preserve integrity), shame (i.e. the emotion experienced when a negative appraisal of the all-inclusive self is concerned, when an individual experiences humiliation or feels as though others find the individual insignificant or worthless, and can result in a momentary inability to think logically and efficiently), and guilt (i.e. associated with negative self-appraisal, but unlike shame, it is related to specific actions and behaviors).

d. **Antagonistic Emotions.** Refers to the class of utilitarian emotions including envy (i.e. the emotion evoked when an individual desires what another individual possesses and feels a sense of inferiority in comparison to the other individual), disgust (i.e. the emotion experienced when an individual is confronting something considered repulsive or abhorrent), contempt (i.e. the emotion experienced when an individual feels superior and dominant, but typically does not engage in aggressive behaviors such as assault), and anger (i.e. the emotion experienced when one feels a sense of wrongdoing, on their part or by other individuals, which is considered offensive and is accompanied by a sense of injustice, unfairness, or inequality).

e. **Aesthetic Emotions.** Refers to the class of emotions including those characteristic of an absence or a less pronounced function for immediate adaptation to a situation of which requires goal relevance evaluation and coping potential. In other words, the aesthetic experience of a work of art or a piece of music is not formed through the appraisal of whether the work meets physical needs, assists in furthering current goals or projects, or is in accordance with personal social values. Instead, aesthetic emotions are the product of an appreciation for the intrinsic qualities of naturalistic beauty, or the qualities of a work of art or artistic performance. A few examples of aesthetic emotions include being moved, in awe, full of wonder, admiration, bliss, ecstasy, fascination, harmony, rapture, and solemnity.

f. **Compound.** Several submodalities or emotions are involved, in which case each submode and identifiable emotion involved should be noted.

Associated Mental Health Findings

Mental health disorders somewhat common in experients of empathic interaction include: Attention Deficit/ Hyperactivity Disorder, Bipolar Disorder, Depressive Disorder, Generalized Anxiety Disorder, Obsessive Compulsive Disorder, and Panic Disorder with or without Agoraphobia, (Kelly, 2012).

Associated Medical Condition Findings

Physical medical conditions somewhat common in experients of empathic interaction can include: Chronic Fatigue Syndrome, Digestive Disorders, and Hypertension or Hypotension (Kelly 2012).

Differential Classification

A wide variety of extrasensory phenomena can present with similar phenomenology. These include:

o **Clairvoyance** and **Telepathy**. Applied when more than emotional content is perceived such as images, sounds, and other sensations. If the experient reports some purely empathic experiences and some experiences involving more than emotional content, the experient should be considered clairvoyant or telepathic with occasional intuitive impressions and/or hallucinations.

o **Empathic Simulation**. Applied when the empathic process involves the experient having a similar emotional experience at the time of an emotional information transfer with a subject (i.e. the experient is "sharing" an emotional experience of their choosing in the subject that will be accommodating to the experients needs/intentions).

Criteria for Empathic Interactive Experiences

A. **Characteristic phenomenology:** all of the following are required criteria for empathic interactive experiences including criteria for empathy in general.

(1) Emotional information is sent by the agent and received by the subject.

(2) Involves one or more individuals and direct emotional information transfer.

(3) Subconscious need for emotional information impression present at the time of the experience.

3.3 Empathic Simulation (ES)

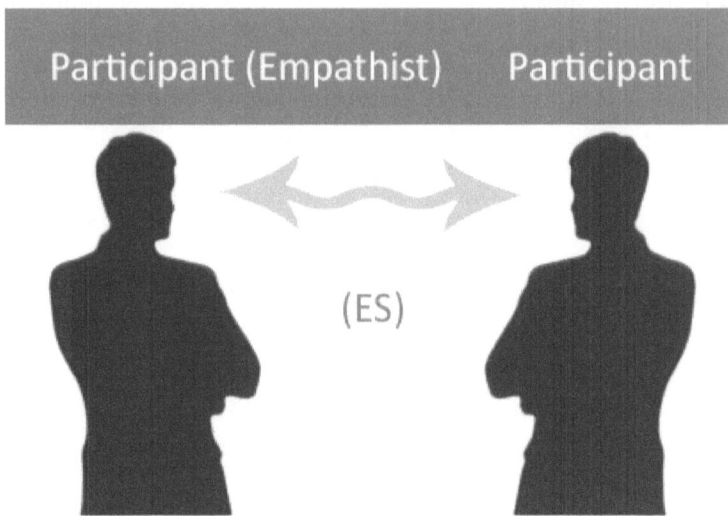

Participant (Empathist)   Participant

(ES)

Phenomenological Features

The essential feature of the (ES) type of empathy is a *case in which an individual's emotional experience appears to directly produce a similar emotional experience in someone else without the intervention of the five senses*. Empathic simulation appears to involve the empathist's emotional experience producing a similar emotional experience in a participant or vice versa. Through this type of empathy, the participant does not "know" empathically what the emotional experience of the participants are, nor is the emotional experience "impressed," but rather it appears that that the emotional experience of the empathist and participants instantaneously become qualitatively identical. The identicalness of the emotional experience is debatable, as there is no empirical evidence to support this at this time. However, reports in regard to this form of tele-empathy suggest exact, or nearly exact, emotional experiences.

This type of empathy also appears to be more non-invasive as participants are typically unaware that, or do not "know" that, the emotional experience is "not their own," as it appears to be less intrusive than empathic cognition or interaction. The skilled empathist would however be able to identify that the simulated emotional experience originated from him/herself if the empathist knowingly shared the emotion with participants. In other words, the empathist can share his/her own emotional experiences with participants, or the empathist can evoke the sharing process of the participant's emotional experience to replace his/her own emotional experience. In the end, I believe the most efficient way to view empathic simulation is as though the emotional experiences have been shared via the exact transmission of the experience from one participant to the other (Kelly, 2012).

Intention Specifiers

The first set of specifiers is for identifying whether the experience was intentional or unintentional.

A. **Spontaneous.** This specifier applies when the empathist and participants share emotional information void of conscious intent on either the empathist's or any of the participant's behalf.

B. **Intentional.** This specifier applies when the empathist intentionally specifies who is to participate in the sharing process, or what emotional information will be shared. If the process involves 'what' rather than 'who,' participants may be selected subconsciously based on their relativity to the required result.

The second set of specifiers is for identifying the subconscious or conscious need or goal that is assumed to be the catalyst for initiating empathic simulative processes.

A. **Adaptive.** This specifier applies when emotional information shared is initiated to assist the participants in understanding and adapting to the empathist's, or group's and the empathist's, needs or goals. The most common goal is to provide emotional comfort and/or a sense of security. Here the empathist and participants typically have some level of emotional investment in each other or the situation in which they occupy.

B. **Directive.** This specifier applies when emotional information shared is initiated to assist the empathist or participants in an action towards a goal (i.e. motivation). In other words, to provide purpose and direction to behavior. This is done in expectation that the empathist and/or participants will be provoked into acting or behaving in a specific or generalized manner as a direct result of the simulated emotional state. Behaviors are <u>affectional</u> (i.e. actions which are taken due to one's emotions, to express personal feelings). However, while emotional information is exclusively involved in empathic simulation, it does not appear to penetrate the barrier of self-control like impression. Here the empathist and participants typically have some level of emotional investment in each other or the situation in which they occupy.

The third set of specifiers is for identifying the direction of the empathic simulative experience.

A. **Input.** This specifier applies when a participant shares emotional information with the empathist. Here emotional information in regard to a participant has been shared with the empathist (e.g. the empathist was feeling anxious, but a participant was not feeling anxious prior to simulation; however post simulation, neither the empathist nor the participant felt anxious).

B. **Output.** This specifier applies when the empathist shares emotional information with a participant. Here emotional information in regard to the empathist has been shared with a participant (e.g. the participant was feeling anxious, but the empathist was not feeling anxious prior to simulation; however post simulation, neither the empathist nor the participant felt anxious).

Course Specifiers

These specifiers are for identifying the characteristic course of empathic simulative experiences over time.

A. **Single Episode.** This specifier applies when the empathist or a participant shares emotional information with the other and the experients reports no prior history of episodes. This specifier also applies when the empathist or participants share emotional information to provide purpose and direction to behavior in which the empathist or a participant reports as not typical (i.e. the empathist or participant has not responded in such a way in similar circumstances in the past). Classification can be very difficult in this case, as it is often difficult to identify if the individual reporting the experience is the empathist or a participant (i.e. if they are the initiator of empathic simulative processes).

B. **Episodic.** This specifier applies when an empathist shares emotional information with other participants of which seems to occur irregularly and of which the duration of the experience is very momentary. An episodic emotional simulation may involve a quick burst of emotion with the duration of the experience lasting only a maximum of a couple of seconds. This specifier also applies when the empathist or a participant shares their emotional state to promote adaptive or directive behavior in the other.

C. **Continuous.** This specifier applies when an empathist or a participant shares emotional information with other of which seems to occur in a continual manner, or when episodes are so frequent it is difficult for either the empathist or the participant to determine where one episode ends and another begins (e.g. prolonged and closely spaced episodes).

Modality Specifiers

This specifier is for identifying the characteristic mode(s) and submode(s) of an empathic simulative experience. In any case, some emotional investment in the participant, or the situation in which the participant resides, on the experient's behalf is expected.

A. **Dream.** Refers to empathic simulation during sleep where the empathist shares information with another participant during the dream state to promote – once awake -- adaptive or directive behavior.

B. **Intuitive Impressions/Emotional.** Refers empathic simulation during normal or altered (e.g. trance) states of conscious awareness, of which can be described as emotional content shared between the empathist and a participant that results in adaptive or directive behavior.

    a. **Achievement Emotions**. Refers to the class of utilitarian emotions including <u>pride</u> (i.e. associated with an enhancement of ego-identity and self-esteem), <u>elation</u> (i.e. provides an individual with the feeling of living fully), <u>joy</u> (i.e. elicits confidence, comfort, and boosts self-esteem), and <u>satisfaction</u> (i.e. is contributing to a feeling of fulfillment and wellbeing).

    b. **Approach Emotions.** Refers to the class of utilitarian emotions including <u>relief</u> (i.e. results subsequent to a negative emotion when an event has taken a turn for the betterment of the individual or group), <u>hope</u> (i.e. contains some level of uncertainty because it is future orientated, but plays a vital role in adaptation as a means to pursue ones goals), <u>interest</u> (i.e. the emotion elicited when one experiences a feeling of engagement, fascination, and curiosity), and <u>surprise</u> (i.e. accompanied by uncertainty, which keeps one on their toes and stimulated as a means to cope with and adjust to new and unexpected actions and events).

    c. **Resignation Emotions.** Refers to the class of utilitarian emotions including <u>sadness</u> (i.e. evoked when one loses something in life such as a loved one, employment, or social standing, and is typically correlated with resignation and failure), <u>fear</u> (i.e. activates a sense of threat, or uncontrollability, alongside a need to preserve integrity), <u>shame</u> (i.e. the emotion experienced when a negative appraisal of the all-inclusive self is concerned, when an individual experiences humiliation or feels as though others find the individual insignificant or worthless, and can result in a momentary inability to think logically and efficiently), and guilt (i.e. associated with negative self-appraisal, but unlike shame, it is related to specific actions and behaviors).

d. **Antagonistic Emotions.** Refers to the class of utilitarian emotions including envy (i.e. the emotion evoked when an individual desires what another individual possesses and feels a sense of inferiority in comparison to the other individual), disgust (i.e. the emotion experienced when an individual is confronting something considered repulsive or abhorrent), contempt (i.e. the emotion experienced when an individual feels superior and dominant, but typically does not engage in aggressive behaviors such as assault), and anger (i.e. the emotion experienced when one feels a sense of wrongdoing, on their part or by other individuals, which is considered offensive and is accompanied by a sense of injustice, unfairness, or inequality).

e. **Aesthetic Emotions.** Refers to the class of emotions including those characteristic of an absence or a less pronounced function for immediate adaptation to a situation of which requires goal relevance evaluation and coping potential. In other words, the aesthetic experience of a work of art or a piece of music is not formed through the appraisal of whether the work meets physical needs, assists in furthering current goals or projects, or is in accordance with personal social values. Instead, aesthetic emotions are the product of an appreciation for the intrinsic qualities of naturalistic beauty, or the qualities of a work of art or artistic performance. A few examples of aesthetic emotions include being moved, in awe, full of wonder, admiration, bliss, ecstasy, fascination, harmony, rapture, and solemnity.

f. **Compound.** Several submodalities or emotions are involved, in which case each submode and identifiable emotion involved should be noted.

Associated Mental Health Findings

Mental health disorders somewhat common in experients of empathic simulation include: Attention Deficit/ Hyperactivity Disorder, Bipolar Disorder, Depressive Disorder, Generalized Anxiety Disorder, Obsessive Compulsive Disorder, and Panic Disorder with or without Agoraphobia, (Kelly, 2012).

Associated Medical Condition Findings

Physical medical conditions somewhat common in experients of empathic simulation can include: Chronic Fatigue Syndrome, Digestive Disorders, and Hypertension or Hypotension (Kelly 2012).

Differential Classification

A wide variety of extrasensory phenomena can present with similar phenomenology. These include:

o **Clairvoyance** and **Telepathy**. Applied when more than emotional content is shared such as images, sounds, and other sensations. If the experient reports some purely empathic experiences and some experiences involving more than emotional content, the experient should be considered clairvoyant or telepathic with occasional intuitive impressions and/or hallucinations.

o **Empathic Cognition.** Applied when the emotional state is "known" rather than "shared." Empathic cognition is often misinterpreted as empathic simulation as both can involve large groups. Differentiating between indirect (i.e. target group » Nature » experient) and direct (i.e. participant » empathist or vice versa) information can be achieved by identifying whether the emotional state was "known" (i.e. the experient was cognizant or aware of the group's emotional state and could identify that the emotional experience was "not their own"), or shared (i.e. the experient experienced the emotional state of the group).

o **Empathic Interaction.** Applied when the empathic process does not involve the empathist having a similar emotional experience at the time of an emotional information empathist-to-subject transfer (i.e. when the empathist is "generating" an emotional experience of their choosing in the subject that will be accommodating to the empathist's needs/intentions).

Criteria for Empathic Simulative Experiences

A. **Characteristic phenomenology:** all of the following are required criteria for empathic simulative experiences including criteria for empathy in general.

  (1) Emotional information is shared between the empathist and one or more participants.

  (2) Involves one or more individuals and direct emotional information sharing.

  (3) Subconscious need for emotional information sharing present at the time of the experience.

GLOSSARY OF TECHNICAL TERMS

A

Agent (*Para-Psychology*)

An individual who sends information using telepathic or empathic processes; who impresses information onto another or is the cause of impression-like telepathic or empathic experience; who initiates experiences involving telepathic or empathic impression; an active and efficient cause; capable of producing a certain effect.

Aha! Factor (*Psychology*)

A subjective standard ascribing validity to an idea when it resonates with one's personal experience. *Similar: Eureka! Effect.*

Alpha Abundance (*Neuroscience*)

An abundance of alpha brain wave activity; an abundance of neural oscillations in the frequency range of 7.5–12.5 Hz and detectable by either electroencephalography (EEG) or magnetoencephalography (MEG).

Altered States of Consciousness (*Psychology*)

Refers to any mental state differing from normal waking conditions. Evidence support these states may be psi-conducive and include; dreaming, hypnosis, hypnogogic-like states induced by the Ganzfeld effect, trance states, meditation, and drug-induced states.

Amygdala (*Neuroanatomy*)

An almond-shaped neural structure in the anterior part of the temporal lobe of the cerebrum; performs a primary role in the processing of memory, decision-making, and emotional reactions.

Angel (*Theology*)

A divine attendant; divine intermediaries; benevolent celestial beings; a spirit once living as human; an intelligent spirit never belonging to the human race and is immortal' protecting and guiding spirits.

Anomalous Communication (*Para-Psychology*)

Communication by paranormal means. A general term referring to unusual experiences involving communications that cannot be explained in terms of current scientific knowledge. *Similar: Psychic communication.*

Apparition (*Para-Psychology*)

An experience usually visual but sometimes involving other sense-modalities in which there appears to be present a person or animal (deceased or living) and even inanimate objects such as carriages and other things, who/which is in fact out of the sensory range of the experient; often associated with spontaneous extrasensory perception, for example, in connection with an agent who is dying or undergoing some other crisis (in which case, it is likely to be termed a "crisis apparition," or in connection with haunting (in which case, it is likely to be referred to in non-technical contexts as a "ghost"). Similar: Ghost; Discarnate Entity, Astral Projection.

Apportation (*Para-Psychology*)

An apport is a physical object which has been paranormally transported into a closed space or container, suggesting the passage of "matter through matter," that is, through intervening solid material objects; the paranormal transference or appearance of an object; the paranormal transference of an article from one place to another, or an appearance of an article from an unknown source that is often associated with poltergeist activity or spiritualistic séances.

Astral Projection (*Metaphysics*)

An astral body is an entity said to be an exact, quasi-physical replica or "double" of the individual physical body, which can separate itself from the physical body, either temporarily, as in dreaming or in the out-of-the-body experience, or permanently, at the moment of death. Also known as the "etheric" body; an interpretation of out-of-body experience (OBE) that assumes the existence of an "astral body" separate from the physical body and capable of traveling outside it; involves a clairvoyant remotely viewing a location joined by the sensation of actually being physically at the location; involves a clairvoyant remotely viewing themselves as in out-of-body experiences; an anomaly of remote viewing/sensing. *Similar: Out-of-Body Experience, Remote Viewing, Apparition.*

Audible Thought (*Psychiatry*)

A person hearing an audible thought in their own head may hear a thought spoken in their own voice; a thought that is perceived as being spoken aloud; hallucinatory voices that echo or speak his thoughts aloud which are associated with Schneider's First Rank symptoms of Schizophrenia. *Similar: Auditory hallucinations.*

Augury (*Metaphysics*)

An event that is experienced as indicating important things to come; a sign of what will happen in the future; an omen; the interpretation of omens; a priest and official in the classical world with the role of interpreting the will of the gods by interpreting omens; prophetic divining of the future by observation of natural phenomena. *Similar: Augur, Divination, Diviner.*

Aura (*Medical Science*)

A perceptual disturbance experienced by some with migraine or seizures before either the headache or seizure begins; the perception of a strange light symptomatic of dehydration or sleep deprivation; visual hallucinations such as bright lights and vibrating visual field often in conjunction with other perceptual disturbances and sensations. *Similar: Hallucination.*

Aura (*Para-Psychology*)

A field of energy believed by some to surround living organisms; a field of subtle, luminous radiation surrounding, emanating from, a person or object; thoughts and feelings surrounding a person in the form of colors and seen by the third eye.

Automatism (*Medical Science*)

Any complex sensory or motor activity the details of which are carried out by a person without their conscious awareness or volition, thus constituting instances of dissociation; examples of sensory automatisms are certain visual and auditory hallucinations; an example of motor automatisms is sleep-walking. A symptom of schizophrenia.

Automatism (*Para-Psychology*)

A sprit's influence on a medium's physical activity (e.g. arm movements, writing, drawing, trance-utterances). *Similar: Automatic writing, Xenoglossy.*

B

Boundary Cases (*Psychiatry*)

Clinical cases that straddle the boundary of two or more diagnoses.

Boundary Cases (*Para-Psychology*)

Extrasensory experience cases that involve the criteria of two or more types or subtypes of extrasensory experience.

C

Chance (*Philosophy*)

A measure of how likely it is that some event will occur; a number expressing the ratio of favorable cases to the whole number of cases possible; an unknown and unpredictable phenomenon that causes an event to result one way rather than another. *Similar: Luck, Probability.*

Chance (*Para-Psychology*)

A shorthand expression used in parapsychology referring to the *Mean Chance Expectation (MCE)*.

Channeling (*Para-Psychology*)

When an entity influences a medium's physical body in addition to speech; a phenomenon where one psychically conveys information or messages from another consciousness other than their own, typically an entity. *Similar: Mediumship, Trance Mediumship, Xenoglossy, Automatism, Clairvoyant Interaction.*

Clairvoyance (*Para-Psychology*)

The psychical influence of an objective environment, which is assumed capable of storing, retaining, and recalling information pertaining to the past and current states of objects and events, involved in the interim integration, processing, shifting, and retrieval of information pertaining to objects and events in real-time, and probabilistically determining the potential trajectory of future events.

Clairvoyant Cognition (*Para-Psychology*)

The phenomenologically indirect knowledge of an object or event via Nature without the intervention of the five senses. This includes remote viewing/sensing, precognition, retro/postcognition, and contemporaneous ESP.

Clairvoyant Interaction (*Para-Psychology*)

The causal influence of an entities "mind" onto an experient without the intervention of the five senses. This form of clairvoyance is typically known as trance mediumship or channeling.

Clairvoyant Simulation (*Para-Psychology*)

A case in which an experient's mental or physical state appears to produce an accommodating effect in Nature, or Nature produces an accommodating effect within itself or the experient to satisfy the needs of the experient. *Similar: Wish Fulfillment, Serendipity, Synchronicity, Luck, Probability Shifting, Prayer Fulfillment, Spell Casting, Intercessory Prayer.*

Contemporaneous ESP (*Para-Psychology*)

Refers to clairvoyant extrasensory experiences in relation to persons, places, objects, or events in real-time. *Similar: Remote Viewing/Sensing.*

Coincidence (*Statistics*)

The temporal property of two things happening at the same time; An event that might have been arranged although it was really accidental; a collection of two or more events or conditions, closely related by time, space, form, or other associations which appear unlikely to bear a relationship as either cause to effect or effects of a shared cause, within the observer's or observers' understanding of what cause can produce what effects. *Similar: Luck, Chance.*

Cryptomnesia (*Psychology*)

A memory bias that occurs when a forgotten memory returns without it being recognized as such by the subject, who believes it is something new and original.

D

Determinism (*Philosophy*)

A philosophical theory holding that all events are inevitable consequences of antecedent sufficient causes; often understood as denying the possibility of free will. *Opposite: Free Will, Indeterminism.*

Diagnostic and Statistical Manual of Mental Disorders (DSM)

The Diagnostic and Statistical Manual of Mental Disorders, published by the American Psychiatric Association, offers a common language and standard criteria for the classification of mental disorders. The current editions is the DSM-5.

Discarnate Spirits/Entity (*Para-Psychology*)

Typically refers to spirits of the deceased or consciousness that is assumed to exist after bodily death; a disembodied being, as opposed to an incarnate one; the surviving personality of a deceased individual or non-human entity; a spirit; ghost; a disembodied consciousness of some type of immaterial or discarnate element of human, or animal, personality that survives bodily death for an unknown duration of time. *Similar: Entity, Ghost, Spirit.*

Distance Healing (*Para-Psychology*)

Refers to instances of psychic healing that occurs at a distance where the healer is typically miles away from the subject such as the healer and subject residing in different buildings, cities, or even countries, and may be the result of remote bio-PK (i.e. remote psychokinetic energy healing) or intention (i.e. intention healing). Can also refer to in-room energy healing that does not require touch such as Reiki. *Similar: Clairvoyant Simulation, Psychic Healing, Psychokinesis.*

Divinatory Practices (*Metaphysics*)

Divination is the practice of seeking knowledge of the future or the unknown by supernatural powers; Divination is the attempt to gain insight and information into a query, person, object, or situation by way of a standardized process. Diviners ascertain their interpretations of how they or a querent should proceed by reading signs, or events. There have been hundreds of types of divinatory methods used and recorded throughout history, but the most common forms today are of the sortilege type in which consists of the casting of lots. Popular types of divination include the I Ching, tarot or oracle cards, runes, and stones. *Similar: Divination, Diviner.*

E

Electroencephalogram (EEG) (*Neuroscience*)

An EEG machine, or electroencephalograph, in an experimental context, refers to the recording of the brain's spontaneous electrical activity over a short period, as recorded from multiple electrodes positioned on the scalp; a device that measures brainwave activity.

Electrogastrogram (EGG) (*Neuroscience*)

A graphic produced by an electrogastrograph, which records the electrical signals that travel through the stomach muscles and control the muscles' contractions.

Emotion (*Social Science*)

An episode of interrelated, synchronized changes in the states of all or most of the five organismic subsystems in response to the evaluation of an external or internal stimulus event as relevant to major concerns of the organism.

Empath (*Para-Psychology*)

A person with the psychic ability to apprehend the emotional state of other individuals in a way currently unexplained by conventional science; an individual that is highly psychically sensitive to the emotions of others; a person who is capable of psychically sensing the emotions of others; a person with frequent extrasensory empathic experiences. *Similar: Empathy, Empathic, Empathetic, Intuitive Impressions.*

Empathy (*Para-Psychology*)

The psychical influence of emotion via experient influence over the emotional basis of consciousness and the mental and physiological processes associated with a wide variety of emotional experiences.

Empathic Cognition (*Para-Psychology*)

The phenomenologically indirect knowledge of the collective emotional experience of a large group or population via Nature.

Empathic Interaction (*Para-Psychology*)

The direct causal influence of an emotional experience of another individual without the intervention of the five senses.

Empathic Interaction *Hypnogenic* (*Para-Psychology*)

A form of empathic interaction that is assumed involved in causing a mild hypnotic state in subjects, possibly involving impressed emotions related to relief, which results in the behaviors of relaxation or decompression, via an empathist's command.

Empathic Simulation (*Para-Psychology*)

A case in which an individual's emotional experience appears to directly produce a similar emotional experience in someone else without the intervention of the five senses.

Energy (*Physics*)

Typically understood as the ability that a system has to do work on other systems. A thermodynamic quantity equivalent to the capacity of a physical system to do work.

Entity (*Para-Psychology*)

A being of non-corporeal (non-physical) existence. A being with distinct and independent existence; a conscious being not of this world; the consciousness of a deceased person; an angel, demon, or other types of supernatural beings including seemingly omnipotent beings. *Alternative: Ghosts, Apparition, Spirit.*

Entity (*Philosophy*)

That which is perceived, known, or inferred to have its own distinct existence (living or nonliving). *Similar: Object.*

Eureka! Effect (*Psychology*)

The common human experience of suddenly understanding a previously incomprehensible problem or concept typically presumed as a result of subconscious intuition; a sudden unexpected discovery; a suddenly realized solution to a previously unconceivable problem; the effect of solving a problem by allowing the subconscious to work while the conscious mind lets go (e.g. during sleep); spontaneous inspiration. *Similar: Aha! Factor, Intuition, Intuitive Impression.*

Experient (*Psychology*)

A person undergoing an experience or that had an experience.

Extrasensory Experiences (*Para-Psychology*)

A single, episodic, or continuous experience of an extrasensory nature. *Similar: Extrasensory Perception.*

Extrasensory Perception (ESP) (*Para-Psychology*)

The reception of information not gained through the recognized physical senses, but sensed by the mind; the acquisition of information about, or response to, an external event, object, or influence (e.g. mental or physical; past, present, or future) otherwise than through any of the known sensory channels. *Similar: Extrasensory Experiences.*

F

Fortune Teller (*Metaphysics*)

A person who foretells the future by paranormal means. *Similar: Psychic, Diviner, Clairvoyant, Precognitive.*

Functional Magnetic Resonance Imaging (fMRI) (*Neuroscience*)

A functional neuroimaging procedure using MRI technology that measures brain activity by detecting associated changes in blood flow.

G

Gaussian Probability Distributions (*Statistics*)

Assumes that observations are closely clustered around a mean, and this amount is decaying quickly the farther away from the mean; a function that tells the probability that any real observation will fall between any two real limits or real numbers, as the curve approaches zero on either side. Also known as normal distribution.

Ghost (*Para-Psychology*)

The apparition of a deceased person, frequently similar in appearance to that person, and usually encountered in places he or she frequented, the place of his or her death, or in association with the person's former belongings. *Similar: Discarnate Entity, Apparition.*

Ghost Whisperer (*Popular Culture*)

Refers to an individual that is capable of passing on and conveying messages from discarnate spirits to the living. *Similar: Mental Mediumship, Trance Mediumship, Channeling.*

Guardian Angel (*Theology*)

A spirit that is believed to watch over and protect a person or place; an angel assigned to protect and guide a particular person, group, kingdom, or country; serves to protect whichever person 'God' assigns them to, and present prayer to 'God' on that person's behalf. *Similar: Angel, Entity.*

Gut Feeling or Gut Intuition (*Psychology*)

A personal, intuitive or instinctive feeling or response; sudden, strong judgments of which the origin cannot be immediately explained. *Similar: Intuition, Intuitive Impression.*

H

Hallucination (Para-Psychology)

An experience involving the apparent perception of someone or something not physically present, that is not indicative of psychopathology, and where veridical information is conveyed. In hallucinatory extrasensory experiences, information is conveyed in the form of a sensory hallucination. Hallucinations are typically reported as being abstract, fragmented, or somewhat vague, or comprehensible and weak, or vivid and substantial and are perceived in a conscious and awake state. Hallucinations can occur in any sensory modality including visual (sight), auditory (hearing), olfaction (smell), gustatory (taste), tactile (pressure/touch), equilibrioception (balance), thermoception (temperature), chronoception (time), etc. Hallucinations often involve living persons or the deceased.

Hallucination (*Psychology*)

An experience involving the apparent perception of something not present; sensing things while awake that appear to be real, but instead have been created by the mind; something (such as an image, a sound, or a smell) that seems real but does not really exist and that is usually caused by mental illness or the effect of a drug.

Hallucinations *Auditory* (*Psychology*)

The perception of sound without outside stimulus; hallucinations of hearing/sound. Involves verbal and non-verbal hallucinations perceived internally or externally either with or without an identifiable location of origin. Auditory hallucinations can be divided into two categories: elementary and complex. Elementary hallucinations are the perception of sounds such as hissing, whistling, an extended tone, and more. Complex hallucinations are those of voices, music, or other sounds that may or may not be clear, may be familiar or completely unfamiliar, and friendly or aggressive, among other possibilities. Similar: Audible Thoughts, Thought Insertion.

Hallucinations *Gustatory* (*Psychology*)

The perception of taste without a stimulus. May include a wide range of taste sensations classified as bitter, sour, sweet, 'disgusting,' etc., but can be classified in more specific terms (e.g. tobacco, garlic, salt, blood, etc.).

Hallucinations *Olfactory* (*Psychology*)

Smelling odors that are not really present. The most common odors are unpleasant smells, typically indicative of psychopathology or pathology, such as rotting flesh, vomit, urine, feces, smoke, or others. These hallucinations are typically extrinsic where the localization of the smell is outside of, or on, the body.

Hallucinations *Somatic* (*Psychology*)

Refers to hallucinations from inside the body only (e.g. heart, lungs, sensations within the limbs, stomach e.g. nausea). Also known as somatosensory hallucinations.

Hallucinations *Tactile* (*Psychology*)

Tactile hallucinations are the illusion of tactile sensory input, simulating various types of pressure to the skin or other organs. Tactile hallucinations are classified based on the type of sensation experience (e.g. painful sensations are classified as pain hallucinations; temperature sensations are classified as thermal/thermic hallucinations).

Hallucinations *Visual* (*Psychology*)

The perception of an external visual stimulus where none exists; halluci-nations of sight. Involving a perceived complexity classified as simple or complex. If the entire environment is replaced by the visual hallucination, the hallucination is classified as scenic or panoramic hallucinations. Visual hallucinations in which are located beyond the visual field (e.g. in the back of the mind, third eye vision, etc.) are classified as extracampine hallucinations. Using the perceived shape of the hallucination, visual hal-lucinations can be classifies as formed, organized, or unformed (i.e. ab-stract).

Haruspex (*Metaphysics*)

In ancient times, a religious official who interpreted omens by inspecting the entrails of sacrificial animals. *Similar: Diviner.*

Haunting (*Para-Psychology*)

The occurrence of paranormal phenomena associated with a particular locality (especially a building) and usually attributed to the activities of a discarnate entity. The phenomena may include apparitions, poltergeist disturbances, cold drafts, sounds of steps and voices, and various odors. *Similar: Discarnate Entity, Poltergeist.*

Hippocampus (*Neuroanatomy*)

A curved elongated ridge that extends over the floor of the descending horn of each lateral ventricle of the brain, that consists of gray matter covered on the ventricular surface with white matter, and that is in-volved in forming, storing, and processing memory.

Holy Spirit (*Theology*)

The third person in the Christian trinity. In Christian theology, the Holy Spirit is believed to perform specific divine functions in the life of the Christian or the church. The Holy Spirit also acts as comforter, one who intercedes, or supports or acts as an advocate, particularly in times of tri-al. *Similar: Holy Ghost, Nature.*

Hypnagogic State (*Psychology*)

A transitional state of consciousness experienced just before falling asleep often characterized by vivid hallucinations. Alternatively used to refer to the similar state during the process of awaking. *Similar: Hypna-gogia, Hypnopomp.*

Hypnopompic State (*Psychology*)

Of or relating to the state immediately preceding waking up; the state of consciousness leading out of sleep. *Similar: Hypnopomp, Hypnagogia.*

Hypnosis (*Psychology*)

A condition or state that resembles sleep but that is induced by suggestion; as state accompanied by narrowing of the range of attention; a temporary state of heightened relaxation and suggestibility during which some (not all) people are able to become so focused that they experience imaginary happenings as if they were real.

I

I Ching (*Metaphysics*)

One of the oldest of the Chinese classic texts. The book contains a divination system. In Western cultures and modern East Asia, it is still widely used for this purpose. *Alternative: Classic of Changes; Book of Changes; Zhouyi. Similar: Divination.*

Implicit Emotional Memories (*Psychology*)

Emotional memories referred to as "non-declarative" because an individual is unable to verbally "declare" these memories. Implicit memories are non-conscious and involve specific feelings/emotions (e.g. ones conditioned emotional response to a "frightening" situation), which is not something that one can consciously or verbally recall.

Information Teleportation *Quantum* (*Para-Psychology*)

Referring to the process of instantaneous information transmission across space-time involving two or more quantum entangled participants (e.g. telepathic information transmission from agent to sender).

Information Teleportation *Quantum* (*Physics*)

A process by which quantum information can be transmitted from one location to another via quantum entanglement between the sending and the receiving.

Insight (*Psychology*)

The capacity to gain an accurate and deep intuitive understanding of a person or thing; the understanding of a specific cause and effect in a specific context; an introspection; the power of acute observation and deduction, penetration, discernment, perception called intellection or noesis.

Intention (*Philosophy*)

An agent's specific purpose in performing an action or series of actions; the thing that one plans to do or achieve; an aim or purpose.

Intention Healing (*Para-Psychology*)

Refers to clairvoyant simulative healing, or psychic healing, where the healer focuses on the intended result rather than the actual healing process or the ailment to be healed. This form of healing primarily refers to information-based healing (i.e. bio-logical/genetic information shifting, where the healer and Nature are the source of the healing process), but may involve bio-energy or energy influences external to the body. *Similar: Clairvoyant Simulation Output, Wish Fulfillment, Prayer Healing, Faith Healing.*

Intuition (*Para-Psychology*)

The capability of "paranormally' coming to an idea directly, by means other than those of reasoning and intellect, and often outside of all conscious processes. *Similar: Intuitive Impression, Intuitive, Empath. Empathic.*

Intuition (*Psychology*)

The capability of coming to an idea directly, by means other than those of reasoning and intellect, and often outside of all conscious processes.

Intuition *Mothers* (*Para-Psychology*)

Suggestive of an empathic or telepathic connection between mother and child often seen in a spiritual context.

Intuitionism (*Philosophy*)

The doctrine that knowledge is acquired primarily by intuition.

Intuitive Impression (*Para-Psychology*)

Cover any extrasensory experience in which comprises of a general unreasoned impression or hunch. With intuitive experiences, there is no visual imagery, or auditory information in accompaniment, nor any conscious processes of linear coherent thought leading to the impression. The experient reports suddenly "just knowing" something, typically describing a vague impression, that upon subsequent investigation was supported to some extent. *Alternative: Empathy, Intuitive Telepathy, Intuitive Clairvoyance.*

K

Kundalini (*Metaphysics*)

An indwelling spiritual energy that can be awakened in order to purify the subtle system and ultimately to bestow the state of Yoga, or divine union upon the seeker of truth. It is reported that Kundalini awakening results in deep meditation, enlightenment, and bliss.

L

Levitation (*Para-Physics*)

The psychical influence of suspending an object against the force of gravity in a stable position void of solid, physical contact. *Alternative: Psychokinesis.*

Luck (*Philosophy*)

An unknown and unpredictable phenomenon that leads to a favorable outcome; An unknown and unpredictable phenomenon that causes an event to result one way rather than another; one's overall circumstances or condition in life. *Similar: Chance, Coincidence.*

M

Magic (*Metaphysics*)

The power of influencing the course of events by using supernatural forces; an attempt to understand, experience, and influence the world using rituals, symbols, actions, gestures and language. *Similar: Magick, Spell Casting, Wish Fulfillment, Prayer Fulfillment, Clairvoyant Simulation, Probability Shifting.*

Medical Intuition (*Para-Psychology*)

An alternative medical practice where a practitioner claims to use their self-described intuitive abilities to find the cause of a physical or emotional condition. *Similar: Clairvoyant, Clairvoyance, Medical Clairvoyant, Medical Psychic, Intuitive Counselor.*

Meditation (*Philosophy*)

A practice in which an individual trains the mind or induces a mode of consciousness, either to realize some benefit or as an end in itself. The term itself is widely utilized, but it is typically utilized in a highly vague manner, whereby rendering its descriptive authority inadequate. However, in regard to meditative techniques in which result in certain alternative states of consciousness, some effects are measureable and repeatable. This suggests that despite certain historical attempts to categorize diverse meditation practices (e.g. involving the assumption that meditation is a mystical experience that transcends thought, language, reason, and normal perceptive processes) reliable neuroscientific examinations of meditation are possible. These examinations involve the effects of meditative practices on the brain and body, and typically involve identifying neuroelectric or neuroimaging correlates. *Similar: Altered States of Consciousness.*

Mediumship *Mental* (*Para-Psychology*)

The anomalous communication with immaterial entities most commonly referred to as discarnate spirits (i.e. ghosts) or spirit guides, which are believed to have a form of consciousness and element of personality, or the anomalous communication with Nature, which some believe to be an aggregate of consciousness, or "universal consciousness," assumed "friendly," but possessing a neutral personality (i.e. behaviors, temperament, emotions, etc.). *Similar: Clairvoyance, Trance Mediumship, Channeling, Clairvoyant Interaction.*

Mediumship *Physical* (*Para-Psychology*)

Refers to a medium with or without obvious psychical ability, which can be used as a source of "power" for spirit manifestations such as loud rapping and other noises, voices, materialized objects, materialized spirit bodies, or body parts such as hands, or the occurrence of levitation; the influence of the energies and energy systems of spirits.

Mediumship *Trance* (*Para-Psychology*)

A form of mediumship that has characteristics of mental mediumship, but with the medium sitting or lying down in a deep meditative state. During these sessions, the medium may speak as though the information is being conveyed *to* the medium, but rather, the information is coming directly from the entity, but being conveyed *through* the mediums natural voice and behaviors. During a session of this type, the mediums consciousness is believed to be "set aside" within their physical body as a means to allow an entity's consciousness to enter the mediums physical body. This form of mediumship, verses mental mediumship, appears to allow a means of more comprehensible communication between sitters and the entity called. *Similar: Automatism, Mediumship, Clairvoyant Interaction.*

Mental Compulsion (*Psychology*)

An uncontrollable impulse to perform an act, often repetitively, as an unconscious mechanism to avoid unacceptable ideas and desires which, by themselves, arouse anxiety. *Similar: Mental Suggestion, Telepathic Interaction.*

Mental Suggestion (*Psychology*)

The psychological process by which one person guides the thoughts, feelings, or behavior of another; the process by which a physical or mental state is influenced by a thought or idea; the act of inducing hypnosis. *Similar: Mental Compulsion, Mind Influence, Telepathic Interaction.*

Meta-analysis (*Statistics*)

Refers to statistical methods for contrasting and combining results from different studies, in the hope of identifying patterns among study results, sources of disagreement among those results, or other interesting relationships that may come to light in the context of multiple studies.

Memory (*Neuroscience*)

The power or process of reproducing or recalling what has been learned and retained especially through associative mechanisms; the store of things learned and retained from an organism's activity or experience as evidenced by modification of structure or behavior or by recall and recognition; the process in which information is encoded, stored, and re-trieved; a diverse set of cognitive capacities by which we retain infor-mation and reconstruct past experiences, usually for present purposes.

Mind Control (*Psychology*)

A theoretical indoctrination process which results in "an impairment of autonomy, an inability to think independently, and a disruption of beliefs and affiliations. The term has been applied to any tactic, psychological or otherwise, which can be seen as subverting an individual's sense of con-trol over his or her own thinking, behavior, emotions, or decision-making. *Similar: Brainwashing, Coercive Persuasion, Thought Control, or Thought Reform, Thought Insertion, Telepathic Interaction, Clairvoyant Interaction.*

Mind/Mental Rape (*Psychology*)

A fairly common description of some symptoms of schizophrenia typical-ly in relation to thought insertion. *Similar: Thought Insertion, Thought With-drawal, Thought Broadcasting.*

Mind Reading (Para-*Psychology*)

A common way of describing telepathic cognitive experiences; the faculty of discerning another's thoughts through extrasensory means of com-munication. *Similar: Telepathy, Telepathic Cognition, Telepathic Simulation.*

Mind Reading (*Psychology*)

The ability to know another person's thoughts without being told what they are typically through a set of techniques used by mentalists, psy-chics, fortune-tellers, mediums and illusionists to determine or express details about another person, often to imply that the reader knows much more about the person than the reader actually does. *Similar: Cold Read-ing, Mentalism.*

Multidimensional Classification (*Psychology*)

Of or involving several dimensions or scales utilized in the process of classification (i.e. the act of distributing things into classes or categories of the same type); the approach to classification that quantifies an indi-vidual's symptoms or other characteristics of interest and represents them with numerical values on one or more scales or continuums, rather than assigning them to a category.

Mystical Experiences (*Psychology*)

In relation to experiences aimed at human transformation, variously defined in different traditions; a religious or spiritual experience based on
mystical communion with an ultimate reality. *Similar: Religious Experience,
Spiritual Experience, Transpersonal Experience, Exceptional Experience.*

N

Nature (*Philosophy*)

The universe or reality as a whole including all phenomena of the physical world, both understood and not yet understood, ranging in scale from
the subatomic to the cosmic, and the classical to the quantum; an objective environment, which is capable of storing, retaining, and recalling information pertaining to the past and current states of objects and
events, and probabilistically determining the potential trajectory of future events; the universe as an intelligent, loving, and nurturing entity.
Similar: Universal Consciousness, Divine, God, Holy Spirit, Holy Ghost.

New Agers (*Popular Culture*)

An eclectic group of cultural attitudes arising in late 20th century Western society that are adapted from those of a variety of ancient and modern cultures, that emphasize beliefs (as reincarnation, holism, pantheism,
and occultism) outside the mainstream, and that advance alternative approaches to spirituality, right living, and health.

Non-Hallucinatory Sensations (*Psychology*)

Refers to sensations that are purely intuitive or emotional. *Similar: Intuitive Impressions, Empathy, Emotions, Feelings.*

O

Obsession (*Theology*)

Refers to when a single entity, such as a discarnate spirit or "evil" presence, that attempts to interact with a living individual on a regular basis
either in proximity to the living individual or in proximity to the living in-

dividual's home (i.e. recurring often over the course of weeks, months, or years). Here the physical manifestations are perceptible to other individuals. In other words, the experient is haunted by an entity. *Similar: Mediumship, Physical Mediumship, Telekinesis, Psychokinesis, Haunting, Poltergeist.*

Occipital Lobe (*Neuroscience*)

One of the four major lobes of the cerebral cortex in the brain of mammals; the visual processing center of the mammalian brain containing most of the anatomical region of the visual cortex.

Oracle (*Metaphysics*)

An authoritative person in ancient times who divine the future; a person or agency considered to interface wise counsel or prophetic predictions or precognition of the future, inspired by the gods. *Similar: Divination, Clairvoyant, Clairvoyance, Precognition, Precognitive, Fortune Teller.*

Oracle Cards (*Philosophy*)

Similar to tarot cards, but are used for divinatory purposes in relation to one's life path. Typically not utilized for mundane matters such as love and money. *Similar: Tarot Cards, Divination.*

Out-of-Body Experiences (OBEs) (*Para-Psychology*)

An experience, either spontaneous or induced, in which one's center of consciousness seems to be in a spatial location outside of one's physical body; in which a person appears to themselves to possess a duplicate body, sometimes connected to the physical body by a "silver cord;" in which one feels themselves to be entirely bodiless. *Similar: Astral Projection, Remote Viewing.*

P

Palm Reading (*Metaphysics*)

The supposed ability to predict someone's future by interpreting the lines on the palms of their hands; the claim of characterization and foretelling the future through the study of the palm. *Similar: Fortune Telling, Divination.*

Parapsychological Association (*Organization*)

The Parapsychological Association is an international professional organization of scientists and scholars engaged in the study of *psi* (or 'psychic') experiences, such as telepathy, clairvoyance, psychokinesis, psychic healing, and precognition. The primary objective of the PA is to
achieve a scientific understanding of these experiences. First estab-
lished in 1957, the PA has been an affiliated organization of the American Association for the Advancement of Science (AAAS) since 1969. The
PA is a non-profit, non-adjudicating organization that endorses no ideo-
logies or beliefs other than the value of rigorous scientific and scholarly
inquiry. (www.parapsych.org).

Parapsychology Foundation (*Organization*)

The Parapsychology Foundation is a not-for-profit foundation which
provides a worldwide forum supporting the scientific investigation of
psychic phenomena. The Foundation gives grants, publishes pamphlets,
monographs, conference proceedings and the *International Journal of
Parapsychology*, hosts the Perspectives Lecture Series, conducts the
Outreach Program, maintains the Eileen J. Garrett Library with its collection of more than 12,000 volumes and 100 periodicals on parapsychology and related topics, and is proud of its quality paperback imprint, *Helix
Press*. (www.parapsychology.org).

Parietal Lobe (*Neuroscience*)

That part of the cerebral cortex in either hemisphere of the brain lying
below the crown of the head; integrates sensory information among
various modalities, including spatial sense and navigation, the main sensory receptive area for the sense of touch in the somatosensory cortex
which is just posterior to the central sulcus in the postcentral gyrus, and
the dorsal stream of the visual system. The major sensory inputs from
the skin (i.e. touch, temperature, and pain receptors) relay through the
thalamus to parietal lobe.

Participant (*Para-Psychology*)

An individual who shares information using telepathic or tele-empathic
processes; who shares or who has a sharing-like telepathic or tele-
empathic experience; who may or may not initiate experiences involving telepathic or tele-empathic information. *Relates to: Telepathic Simulation, Empathic Simulation.*

Perception (*Psychology*)

The organization, identification, and interpretation of sensory information in order to represent and understand the environment; a basic component in the formation of a concept.

Percipient (*Para-Psychology*)

An individual, who initiates and receives information using telepathic, clairvoyant, or empathic processes; who perceives or who has a perception-like telepathic, clairvoyant, or empathic experience; who initiates experiences involving telepathic, clairvoyant, or empathic information. *Relates to: Telepathic Cognition, Clairvoyant Cognition, Empathic Cognition.*

Photic Stimulation (*Medical Science*)

Intermittent Photic Stimulation, or IPS, is a form of visual stimulation used in conjunction with electroencephalography to investigate anomalous brain activity triggered by specific visual stimuli, such as flashing lights or patterns.

Physical Object (*Physics*)

A collection of masses or smaller physical bodies; a whole or single "thing."

Poltergeist (*Para-Psychology*)

A disturbance characterized by bizarre physical effects of paranormal origin, other than psychokinesis, suggesting mischievous or destructive intent (e.g. unexplained movement or breakage of objects, loud raps, the lighting of fires, and occasionally personal injury to people); spirit phenomena often depending upon the presence of a particular living individual, called the "focus," frequently an adolescent or child, and apparitions are rarely seen; denotes a demonic spirit or malevolent ghost that manifests itself by moving and influencing objects and may or may not have demonic assistance. *Similar: Obsession, Haunting, Ghost, Entity.*

Possession (*Theology*)

The spirit possession of an individual by a malevolent preternatural being, commonly known as a demon. Descriptions of demonic possessions often include erased memories or personalities, convulsions, "fits" and fainting as if one were dying. Other descriptions include access to hidden knowledge (i.e. gnosis) and foreign languages (i.e. xenoglossia), drastic changes in vocal intonation and facial structure, the sudden appearance of injuries (e.g. scratches, bite marks) or lesions, and superhuman strength. Unlike in channeling, the subject has no control over the possessing entity and so it will persist until forced to leave the victim, usually through a form of exorcism.

Postcognition (*Para-Psychology*)

The purported paranormal transfer of information about an event or object in the past; the psychical influence of an objective environment assumed capable of storing, retaining, and recalling information pertaining to the past states of objects, and events, possibly achievable through the act of an experient requesting and receiving information pertaining to past events via Natures "long-term memory" or "long-term information storage." Alternative: Retrocognition. Opposite: Precognition.

Prayer (*Theology*)

An invocation or act that seeks to activate a rapport with a deity, an object of worship, or a spiritual entity through deliberate communication. Prayer can be a form of religious practice, may be either individual or communal, and take place in public or in private. It may involve the use of words or song. When language is used, prayer may take the form of a hymn, incantation, formal creed, or a spontaneous utterance in the praying person.

Prayer Fulfillment (*Theology*)

Refers to when a prayer is answered. *Similar: Wish Fulfillment, Magic, Miracle, Accommodating Effect (CS).*

Prayer Healing (*Theology*)

Healing through prayer; healing effects initiated by prayer. *Similar: Faith Healing, Psychic Healing, Laying on of Hands, Clairvoyant Simulation (Healing), Accommodating Effect (CS), Magic, Miracle.*

Precognition (*Para-Psychology*)

Literally means, "before or prior to knowing," and is defined as a form of extrasensory perception wherein a person is said to perceive information about places or events through paranormal means before they happen; the psychical influence of an objective assumed capable of probabilistically determining the potential trajectory of future events; the extrasensory perception of information pertaining to probability (i.e. the future states of individuals, objects, and events). *Similar: Presentiment, Premonition, Prediction. Opposite: Postcognition, Retrocognition.*

Precognitive Telepathy (*Para-Psychology*)

A classification that typically refers to clairvoyant cognitive processes rather than telepathic; the phenomenologically indirect knowledge of another person's future thoughts or mental states without the intervention of the five senses; the transfer of information, through Psi, about the future of an individual's mind to another individual. *Similar: Clairvoyance, Clairvoyant Cognition, Precognition.*

Premonition (*Para-Psychology*)

Refers to occurrences when an experient of precognition receives information pertaining to future events that are perceived as only emotions (i.e. intuitive impressions); the paranormal acquisition of information concerning the future mental state of another conscious being; A feeling or impression that something is about to occur, often threatening or grim, but no additional information is available (e.g. visual or auditory hallucinations). *Alternative: Presentiment.*

Presentation *Experiential* (*Psychology*)

Refers to the typical characteristics (e.g. evoking negative or positive emotions) associated with a particular type of experience, where the proper interpretation of the experiential presentation often leads to specific categorization.

Presentation *Clinical* (*Medical Science*)

Refers to the typical signs or symptoms that are associated with a particular disease process (e.g. flat affect in schizophrenia). The proper interpretation of the clinical presentation often leads to a specific diagnosis; the constellation of physical signs or symptoms associated with a particular morbid process.

Presentiment (*Para-Psychology*)

Refers to occurrences when an experient of precognition receives information pertaining to future events that are perceived as only emotions (i.e. intuitive impressions); an intuitive feeling about the future. *Alternative: Premonition.*

Pre-stimulus (*Medical Science*)

Also known as a "presponse," refers to physiological activity before a stimulus. *Similar: Presentiment.*

Principle of Causality (*Philosophy*)

States that a cause must precede its effect. In other words, the cause and its effect are separated by a time-like interval and the effect belongs to the future of its cause. This principle therefore states that it is physically impossible for and effect (future) to occur before its cause (present). Therefore, it is physically impossible for information to travel from the future back in time to the present. *Alternative: Causality; Arrow of Time, Indeterminism.*

Probability (*Statistics*)

The extent to which something is probable; the likelihood of something happening or being the case; a probable or the most probable event; the measure of the likeliness that an event will occur. *Similar: Chance, Luck.*

Probability Shifting (*Para-Psychology*)

The psychical influence on Nature's stored probabilistic information (i.e. altering the probabilistic information pertaining to an object or event). Probabilistic information that can be shifted can include historical (also known as retro-PK), real-time, or future probabilities. In regard to all of these types of influence, quantum information is neither created, copied, hidden, nor destroyed, but rather appears to be negotiated (e.g. an accommodating existing future potential is selected rather than various existing non-accommodating potentials). *Similar: Clairvoyant Simulation Output, Psychokinesis, Retro-Psychokinesis, Retro-PK, Historical Shifting.*

Prophet (*Theology*)

An authoritative person who divines the future; someone who speaks by divine inspiration; someone who is an interpreter of the will of God. *Similar: Prophesier, Fortune Teller, Diviner, Divination, Clairvoyant, Precognition.*

Psi (*Para-Psychology*)

An all-encompassing term for extrasensory perception and psychokinesis combined. The term "Psi-Gamma" is typically used to refer to ESP, and "Psi-Kappa" is the term typically used for PK.

Psi-Conducive (*Para-Psychology*)

Advantageous to the occurrence of psi. Tending to bring about or facilitate psi; being partly responsible for psi processes.

Psipathy (*Para-Psychology*)

Refers to psi related ability including telepathy and/or psychic empathy; refers to telepathic or clairvoyant cognitive experiences where only visual information is received. *Similar: Telepathy, Empathy, Clairvoyant Cognition, Telepathic Cognition, Psipathic.*

Psychic Communication (*Para-Psychology*)

Communication by paranormal means. *Similar: Anomalous Communication, Telepathy, Clairvoyance, Mediumship, Empathy.*

Psychic Healing (*Para-Psychology*)

Is the treatment of mental and/or physical illnesses by spiritual means such as crystals, bioenergy work, faith, or prayer; healing by paranormal means. *Similar: Clairvoyant Simulation Output (Healing), Psychokinesis, Bio-Psychokinesis, Bio-PK, Psychic Energy Healing, Direct Mental Influence of Living Systems (DMILS), Prayer Healing, Faith Healing, Intention Healing.*

Psychic Knowledge (*Para-Psychology*)

The gaining of knowledge by paranormal means. *Alternative: Telepathic Cognition/Simulation, Clairvoyant Cognition/Simulation.*

Psychokinesis (PK) (*Para-Physics*)

Is defined as the direct influence of mind on a physical system that cannot be entirely accounted for by the mediation of any known physical energy. PK includes any direct mind over matter interaction or mental influence upon the structure of a physical system; an umbrella term for all paranormally-based kinetic phenomena.

Psychosis (*Medical Science*)

A severe mental disorder in which thought and emotions are so impaired that contact is lost with external reality; refers to an abnormal condition of the mind, and is a generic psychiatric term for a mental state often described as involving a "loss of contact with reality." People experiencing psychosis may exhibit some personality changes and thought disorder. Depending on its severity, this may be accompanied by unusual or bizarre behavior, as well as difficulty with social interaction and impairment in carrying out daily life activities. *Similar: Psychotic, Psychotic Disorder, Psychopathy, Hallucinations, Delusions.*

Psychometry (*Para-Psychology*)

Is assumed a form of retrocognition, and is defined as the psychical ability to make relevant associations from an object of unknown history by making physical contact with that object. *Similar: Clairvoyance, Clairvoyant Cognition.*

Psychopathology (*Medical Science*)

The scientific study of mental disorders, including efforts to understand their genetic, biological, psychological, and social causes; the study of psychological and behavioral dysfunction occurring in mental disorder or in social disorganization.

Psychopompic Activity (*Para-Psychology*)

Typically refers to an experient of mediumship in which is capable of assisting spirits, which have or have not crossed-over, resolve unfinished issues with the living, and where the experient can also assist spirits that have not crossed-over on their journey to the afterlife; a mediumship ability to assist discarnate spirits, that have not crossed-over to the afterlife, to the afterlife. *Similar: Medium, Mediumship.*

R

Realistic Dreams (*Para-Psychology*)

Dreams that are not necessarily completely literal in context, but information does not include the quality of being conveyed in a disguised form. *Opposite: Unrealistic Dreams.*

Remote Viewing (*Para-Psychology*)

The practice of seeking extrasensory impressions (typically visual content) about a distant, spatially or temporally, or unseen target using subjective means. Typically refers to a contemporaneous temporal facet in which appears to involve the perception of information about places or events through paranormal means during the time at which they are occurring. In regard to contemporaneous clairvoyance -- the psychical influence of an objective environment, which I assume is involved in the interim integration, processing, shifting, and retrieval of information pertaining to remote objects and events in real-time. *Related: Clairvoyant Cognition (Visual), Telepathy (Visual), Contemporaneous ESP (Visual). Related: Remote Sensing.*

Retrocognition (*Para-Psychology*)

The purported paranormal transfer of information about an event or object in the past. The psychical influence of an objective environment assumed capable of storing, retaining, and recalling information pertaining to the past states of objects, and events. Presumed achievable through the act of an experient requesting and receiving information pertaining to past events via Natures "long-term memory" or "long-term information storage;" the extrasensory transfer of information about an individual, object, or event in the past; the extrasensory reception of information pertaining to the past states of individuals, objects, and events. *Alternative: Postcognition. Opposite: Precognition.*

Runes (*Philosophy*)

Any character from an ancient Germanic alphabet used in Scandinavia from the 3rd century to the Middle Ages. Also utilized today as a form of divination.

S

Second Sight (*Para-Psychology*)

A form of extrasensory perception, the supposed power to perceive things that are not present to the senses, whereby a person perceives information e.g. in the form of a vision, about future events before they happen, or about things or events at remote locations. *Similar: Telepathic Cognition (Visual), Clairvoyant Cognition (Visual), Clairvoyant Simulation Input (Visual), Remote Viewing, Mediumship.*

Seer (*Metaphysics*)

A person with unusual powers of foresight; an authoritative person who divines the future. *Similar: Diviner, Fortune Teller.*

Sensitive (*Para-Psychology*)

Typically refers to a person who is highly sensitive to the paranormal and paranormal events.

Sensation (*Medical Science*)

A stage of processing of the senses in human and animal systems, such as vision, auditory, vestibular, and pain senses. The faculty through which the external world is apprehended.

Serendipity (*Philosophy*)

Good luck in making unexpected and fortunate discoveries; the occurrence and development of events by chance in a happy or beneficial way. *Similar: Luck, Chance, Coincidence.*

Shaman (*Philosophy*)

A practitioner reaching altered states of consciousness in order to encounter and interact with the spirit world; a person regarded as having access to, and influence in, the world of benevolent and malevolent spirits, who typically enters into a trance state during a ritual, and practices divination and healing. *Similar: Shamanic.*

Sibyl (*Metaphysics*)

In ancient times, a woman who was regarded as an oracle or prophet. *Similar: Diviner, Oracle, Prophet.*

Sitter (*Para-Psychology*)

Refers to any living individual of whom attends a séance or mediumship session that is not the medium. A living individual who sits with a medium at a sitting or séance and who receives information through the medium (i.e. an individual who seeks to sit down with a medium in hopes that the medium can convey information to and/or from a particular discarnate spirit).

Skin Conductance Resistance (*Medical Science*)

A method of measuring the electrical conductance of the skin, which varies with its moisture level. This is of interest because the sweat glands are controlled by the sympathetic nervous system so skin conductance is used as an indication of psychological or physiological arousal. There has been a long history of electrodermal activity research, most of it dealing with spontaneous fluctuations or reactions to stimuli. *Alternative: Skin conductance, also known as galvanic skin response (GSR), electrodermal response (EDR), psychogalvanic reflex (PGR), skin conductance response (SCR) or skin conductance level (SCL).*

Soothsayer (*Metaphysics*)

Someone who makes predictions of the future usually on the basis of special knowledge. *Similar: Oracle, Fortune Teller, Haruspex, Diviner.*

Spell Casting (*Metaphysics*)

A set of words, spoken or unspoken, that is considered by its user to invoke some magical effect and of which may also involve symbols, patterns, recipes, or rituals that are considered by the user as fundamental or supplementary to a magical effect. *Similar: Prayer, Invocation, Magic, Clairvoyant Simulation Output.*

Spirit (*Metaphysics*)

The nonphysical part of a person that is the seat of emotions and character; the soul; the force within a person that is believed to give the body life, energy, and power; the inner quality or nature of a person; That which survives bodily death; consciousness. *Similar: Soul, Consciousness, Ghost, Entity.*

Spirit Guide (*Metaphysics*)

A term typically used by mediums and psychics to describe an entity that remains a discarnate spirit in order to act as a spiritual counselor or protector to one or more living incarnated human beings; refers to an "angelic" being (i.e. of an angelic race) that watches over an individual and protects them, but is not believed to have ever lived and died as a human being. *Similar: Guardian Angel, Holy Spirit, Holy Ghost, Nature, Discarnate Spirit.*

Spiritualism (*Theology*)

A system of belief or religious practice based on supposed communication with the spirits of the dead, especially through mediums; a belief that spirits of the dead have both the ability and the inclination to communicate with the living. *Similar: Spiritualist.*

Spontaneous Experiences (*Psychology*)

Refers to experiences in which are non-consciously controlled and initiated by ones subconscious; occurring as a result of a sudden inner impulse or inclination and without premeditation or external stimulus; occurring without apparent external cause.

Subject (*Science*)

A person or animal that is the object of medical or scientific study; a person likely or prone to be affected by a particular condition or occurrence. *Opposite: Experimenter, Agent.*

Subject (*Para-Psychology*)

An individual, who is influenced into sending or receiving telepathic or tele-empathic information; who telepathic or tele-empathic information is retrieved from, or telepathic or tele-empathic information is impressed upon telepathically or tele-empathically; who is the non-initiator of experiences involving telepathic or tele-empathic information. *Opposite: Agent.*

Synchronicity (*Philosophy*)

Refers to synchronistic events; meaningful coincidences of two apparently (in terms of cause and effect) non-related events. *Similar: Chance, Coincidence, Clairvoyant Simulation Output, Wish Fulfillment, Prayer Fulfillment.*

T

Target (*Para-Psychology*)

In an extrasensory perception test, the person, object, or event that a subject attempts to identify through information extrasensorially acquired, simulated, impressed, etc. In a psychokinesis test, the physical system, or a prescribed outcome (e/.g shifted probability), that the subject attempts to influence or produce.

Tarot Cards (*Metaphysics*)

Any of a set of (usually 72) cards that include 22 cards representing virtues and vices and death and fortune etc.; used for divinatory purposes of typically mundane earthly matters such as love and money. *Similar: Oracle Cards.*

Telepathy (*Para-Psychology*)

The psychical influence of thought via experient influence over the biological basis of consciousness and the mental process by which we perceive, act, learn, and remember. *Similar: Telepathic Cognition, Telepathic Interaction, Telepathic Simulation.*

Telepathy *Dream* (*Para-Psychology*)

Refers to a dream involving a telepathic experience.

Telepathy *Telephone* (*Para-Psychology*)

Refers to a type of telepathy involving an individual determining who was calling another after the call was made, but before the caller spoke.

Telepathy *Twin* (*Para-Psychology*)

Refers to telepathic experience occurring between identical or fraternal twins.

Telepathic Cognition (*Para-Psychology*)

The phenomenologically direct knowledge of another person's thoughts or mental states without the intervention of the five senses.

Telepathic Interaction (*Para-Psychology*)

The causal influence of one mind on another without the intervention of the five senses.

Telepathic Simulation (*Para-Psychology*)

A case in which an individual's mental state appears to produce a similar mental state in someone else without the intervention of the five senses.

Temporal Lobe (*Neuroscience*)

Each of the paired lobes of the brain lying beneath the temples, including areas concerned with the understanding of speech. *Temporal Lobe Dysfunction* refers to damage of the temporal lobe. Individuals who suffer from medial temporal lobe damage have a difficult time recalling visual stimuli. This neurotransmission deficit is due to, not lacking perception of visual stimuli but to, lacking perception of interpretation. The most common symptom of inferior temporal lobe damage is visual agnosia, which involves impairment in the identification of familiar objects. Damage specifically to the anterior portion of the left temporal lobe can cause savant syndrome (i.e. a condition in which a person with a mental disability, such as an autism spectrum disorder, demonstrates profound and prodigious capacities or abilities far in excess of what would be considered normal).

Thought Broadcasting (*Medical Science*)

A delusional belief that others can hear or are aware of an individual's thoughts. It is one of Schneider's First Rank symptoms of schizophrenia. This differs from telepathy in that the thoughts being broadcast are thought to be available to anybody (i.e. strangers), not just those who one has a strong emotional connection.

Thought Insertion (*Medical Science*)

The feeling that one's thoughts are not their own; the person experiencing the thought insertion will not necessarily know where the thought is coming from. It is one of Schneider's First Rank symptoms of schizophrenia. This differs from telepathic interaction in that in cases of TI the "attacker" is identifiable; there is one attacker and one or more victims; the victim and the attacker have or have had a social/romantic emotional connection; the attacker is trying to direct the victim to think, feel, or act a certain way that is beneficial to the attacker.

Thought Transmission (*Para-Psychology*)

Refers to the telepathic process of one individual sending and one individual receiving information telepathically. *Similar: Thought Transference, Thought Reception.*

Thought Withdrawal (*Medical Science*)

The delusional belief that thoughts have been 'taken out' of the patient's mind, and the patient has no power over this. It often accompanies thought blocking (i.e. a phenomenon that occurs in people with psychiatric illnesses (usually schizophrenia), occurs when a person's speech is suddenly interrupted by silences that may last a few seconds to a minute or longer). The patient may experience a break in the flow of their thoughts, believing that the missing thoughts have been withdrawn from their mind by some outside agency. . It is one of Schneider's First Rank symptoms of schizophrenia. This differs from telepathy in that "mind reading," or telepathic cognition, involves "knowing" what one is thinking rather than "stealing their thoughts."

Trance (*Para-Psychology*)

A state of dissociation in which the individual is unaware of their condition and surroundings, and in which various forms of automatism (e.g. automatic writing) may be expressed; typically demonstrated under hypnotic, or mediumistic conditions.

Trance (*Psychology*)

Denotes a variety of processes, ecstasy, techniques, modalities, and states of mind, awareness, and consciousness. May be associated with hypnosis, meditation, magic, flow, and prayer. In addition, may related to the generic term altered states of consciousness.

U

Unrealistic Dreams (*Para-Psychology*)

Are dreams defined similar to realistic dream experiences, but here the imagery is more dramatized by fantasy. Important elements within the spectrum may appear realistic, but the scene or scenarios surrounding the elements appear disguised, as in symbolic form. *Opposite: Realistic Dreams.*

V

Validation (*Para-Psychology*)

Refers to when an extrasensory experience has been validated by an individual other than the experient (e.g. the subject(s) confirmed the accuracy of the information received by the experient), and/or a clinician determines the experience was more than a coincidence/chance occurrence based on the quality of the information received and reported, and all other possible explanations for obtaining the information is excluded.

Visions (*Para-Psychology*)

An experience of seeing someone or something in a dream or trance, or as a supernatural apparition. *Similar: Clairvoyant (Visual), Medium (Visual), Visionary.*

W

Wicca (*Theology*)

The religious cult of modern witchcraft, especially an initiatory tradition founded in England in the mid-20th century and claiming its origins in pre-Christian pagan religions. Wicca is a diverse religion with no central authority or figure defining it. It is divided into various lineages and denominations, referred to as *traditions*, each with its own organizational structure and level of centralization. *Similar: Wiccan Practitioner.*

Wish Fulfillment (*Philosophy*)

The satisfying of unconscious desires in dreams or fantasies. *Similar: Prayer Fulfillment, Clairvoyant Simulation Output, Synchronicity.*

X

Xenoglossy (*Para-Psychology*)

When an entity impresses its own personality or skills onto an experient of which the experient himself/herself does not actually possess (e.g. communicates in a language or vocabulary unknown to the experient). In cases involving Xenoglossy and the spirit of a discarnate entity, the personality or skills of the spirit, from when he/she was alive, are impressed onto the medium for the duration of the session. *Similar: Trance Mediumship, Channeling, Clairvoyant Interaction.*

Y
Z

REFERENCES

Introduction

Barrett, T.R. & Etheridge, J.B. (1992). Verbal hallucinations in normals, I: People who hear 'voices." *Applied Cognitive Psychology, 6, 379-387.*

Cardeña, E., Lynn, S. J., & Krippner, S. (Eds.). (2000). *Varieties of anomalous experience: Examining the scientific evidence.* Washington, DC: American Psychological Association.

Coelho, C., Tierney, I., & Lamont, P. (2008). Contacts by distressed individuals to UK parapsychology and anomalous experience academic research units---A retrospective survey looking to the future. *European Journal of Parapsychology, 23,* 31-59.

Grimby, A. (1998). Hallucinations following the loss of a spouse: Common and normal events among the elderly. *Journal of Clinical Geropsychology, 4, 65-74.*

IGPP. (2007). Counseling and Help for People with Unusual Experiences at the Outpatient Clinic (Ambulanz) of the Psychological Institute at the University of Freiburg. Retrieved from: (www.igpp.de)

Kelly, T.M. (2011a). *Telepathy: A Quantum Approach – The Psychical Influence of Thought.* Charleston SC: Lulu Enterprises Inc.

Kelly, T.M. (2011b). *Clairvoyance: A Quantum Approach – The Psychical Influence of Information and Communication with Immaterial Entities.* Charleston SC: Lulu Enterprises Inc.

Lukoff, D. (2000). DSM-IV Religious and Spiritual Problems. Coursebook. Retrieved from: (www.experiencers.com)

Posey, T. B. & Losch, M. E. (1983). Auditory hallucinations of hearing voices in 375 normal subjects. *Imagination, Cognition and Personality, 3, 99-113.*

Poulton R., Caspi A., Moffit T. E., Cannon M., Murray R., Harrington H. L. (2000). Children's self-reported psychotic symptoms and adult schizophreniform disorder. A 15 year longitudinal study. *Arch. Gen. Psychiatry 57, 1053–1058.*

Romme, M.A.J., Escher, A.D.M.A.C. (1989). Hearing Voices. *Schizophrenia Bulletin, 15, 209-216.*

Shuchter, SR, Zisook, S. (1993). The course of normal grief. In: Stroebe, MS., Stroebe, W., Hansson, RO., eds.: *Handbook of Bereavement: Theory, Research, and Intervention.* Cambridge, United Kingdom: Cambridge University Press, 23-43.

Sidgewick, H., Johnson, A., Myers, F.W.H., et al (1894), Report of the census of hallucinations. *Proceedings of the Society for Psychical Research, 26, 259 - 394.*

Tien, A. Y. (1991). Distributions of hallucinations in the population. *Social Psychiatry and Psychiatric Epidemiology, 26, 287 -292.*

Ullman, M. (1977). *Psychopathology and Psi Phenomena.* Handbook of Parapsychology. Van Nostrand Reinhold Company.

VERDOUX, H et al. (1998). A survey of delusional ideation in primary care patients. *Psychological Medicine, 28, 127–134.*

West, D. J. (1948). A mass-observation questionnaire on hallucinations. *Journal of the Society for Psychical Research, 34, 187–196.*

Core Sections

Beischel, J., & Schwartz, G. E. (2007). Anomalous information reception by research mediums demonstrated using a novel triple-blind protocol. *EXPLORE: The Journal of Science & Healing, 3, 23-27.*

Blackmore, S. (1980). A Study of Memory and ESP in Young Children. *Journal of the Society for Psychical Research, 50, 501-520.*

Braud, W. (2011). Toward More Subtle Awareness: Meanings, Implications, and Possible New Directions for Psi Research. *Mindfield, Parapsychological Association, 3, 1-8.*

Braude, S. (1978). Telepathy. *Noûs 12, 3, 267-301.*

Dalton, K., Utts, J. (1995). Sex pairings, target type and geomagnetism in the PRL automated ganzfeld series. *Proceedings of Presented Papers: The Parapsychological Association 38th Annual Convention, 99-109.*

Haraldsson, E., Houtkooper , J.M. (1991). Psychic Experiences in the Multinational Human Values Study: Who Reports Them? *The Journal of the American Association for Psychical Research, 85, 145-165.*

Irwin, H., Watt, C. (2007). *An Introduction to Parapsychology.* McFarland & Company, Inc.

Kelly, T.M. (2011a). *Telepathy: A Quantum Approach – The Psy-*

chical Influence of Thought. Charleston SC: Lulu Enterprises Inc.

Kelly, T.M. (2011b). *Clairvoyance: A Quantum Approach – The Psychical Influence of Information and Communication with Immaterial Entities.* Charleston SC: Lulu Enterprises Inc.

Kelly, T.M. (2012). *Empathy: A Quantum Approach – The Psychical Influence of Emotions.* Charleston SC: Lulu Enterprises Inc.

Kolodziejzyk, G. (2012). Greg Kolodziejzyk's 13-Year associative remote viewing experiment results. *The Journal of Parapsychology, 76, 349-368.*

McCraty et al., (2004). Electrophysiological evidence of intuition: Part 2: A system-wide process? *The* Journal of Alternative and Complementary Medicine, *10, 325-336.*

Myers, F.W.H. (1903). *Human personality and its survival of bodily death.* London: Longmans, Green.

Palmer, J. (1978). Extrasensory perception: Research findings. In S. Krippner (Ed.), *Advances in Parapsychological Research, Vol 2: Extrasensory Perception,* 144-148. New York: Plenum Press.

Parapsychological Association (1997). Glossary from Thalbourne, M. (2003). Glossary of Terms Used in *Parapsychology.* Puente Publications. Retrieved from: (www.parapsych.org)

Parra, A. (2011). Online Survey of Experiences: Anomalous / paranormal and its emotional impact: relating to gender, age and other variables. *E-Newsletter PSI, 6, 2, Institute of Parapsychology.*

Radin, D. (1997). *The Conscious Universe.: The Scientific Truth of Psychic Phenomena.* HarperOne.

Radin, D. (2000). Time-reversed human experience: Experimental evidence and implications. *Esalen Center for Theory and Research "Subtle Energies and Uncharted Realms of the Mind": an Esalen Invitational Conference.*

Radin, D. (2004). Event-related electroencephalographic correlations between isolated human subjects. *The Journal of Alternative and Complimentary Medicine, 10, 315-323.*

Radin, D., & Schlitz, M. (2005). Gut feelings, intuition, and emotions: An exploratory study. *The Journal of Alternative and Compleme, ntary Medicine, 11, 85-91.*

Radin, D. (2006). *Entangled Minds: Extrasensory Experiences in a Quantum Reality.* Paraview Pocket Books.

Rhine, L. E. (1951). Conviction and associated conditions in spontaneous cases. *Journal of Parapsychology, 15, 164-91.*

Sartori, L., Massaccesi, S., Martinelli, M., & Tressoldi, P.E. (2004). Physiological correlates of ESP: Heart rate differences between targets and nontargets in clairvoyance and precognition forced choice tasks. *The Parapsychological Association Convention, Proceedings of Presented Papers, 407-412.*

Sheldrake, R., Smart, P. (2003). Experimental tests for telephone telepathy. Journal of the Society for Psychical Research, 67, 184-199.

Spotswood, S.J.P. (1990). Geomagnetic activity and anomalous cognition: A preliminary report of new evidence. Subtle Energies, 1, 91-102.

Thalbourne, MA (2001). Broken marital relationships and occurrence of psi phenomena. *Argentina Journal of Parapsychology, 12, 171-175.*

Thalbourne, M. A. (2003). *A Glossary of Terms Used In Parapsychology.* New York: Puente Publications.

Williams, B. J., & Roll, W. G. (2008). Neuropsychological Correlates of Psi Phenomena. *The Joint Annual Convention of The Parapsychological Association, Inc. (51st) and The Incorporated Society for Psychical Research (32nd).*

Zingrone, NL, & Alvarado, CS (1997). "Broken" marital relations and claims of parapsychological experiences. *Paper presented at the 40th Annual Convention of the Parapsychological Association. Brighton, UK.*

Glossary

Barcan, R. (2009). Intuition and Reason in the New Age: A Cultural Study of Medical Clairvoyance. In Howes, David (Eds.), *The Sixth Sense Reader*, (pp. 209-232). Oxford, United Kingdom: Berg Publishers.

Bolte, S. (2004). Comparing the intelligence profiles of savant and nonsavant individuals with autistic disorder. *Intelligence, 32, 121.*

Braud, W. (2000). Toward More Subtle Awareness: Meanings, Implications, and Possible New Directions for Psi Research. *In-*

stitute of Transpersonal Psychology.

Hall, R. (1998). Explicit and Implicit Memory. *Missouri University of Science and Technology.*

Lutz et. al; Slagter, H.A., Dunne, J.D., Davidson, R.J. (2008). Attention regulation and monitoring in meditation. *Trends in Cognitive Sciences, 12, 163-169.*

Oxford Dictionary. (n.d.) United States Version. Retrieved from: (www.oxforddictionaries.com)

Psychology Dictionary. (n.d.) Retrieved from: (www.psychologydictionary.org)

Rai, U.C (1993). *Medical Science Enlightened.* New Delhi: Life Eternal Trust.

Merriam-Webster Online: Dictionary and Thesaurus. (n.d.) Retrieved from: (www.merriam-webster.com)

Miller, L.K. (1999). The savant syndrome: Intellectual impairment and exceptional skill. *Psychological Bulletin, 125, 31–46.*

Morgan, A. (1999). Sahaja Yoga: an ancient path to modern mental health? Doctoral Thesis: University of Plymouth. Retrieved from: (pearl.plymouth.ac.uk)

Parapsychological Association (2014). Organizational Description. Retrieved From: (www.parapsych.org)

Parapsychological Association (1997). Glossary from Thalbourne, M. (2003). Glossary of Terms Used in *Parapsychology.* Puente Publications. Retrieved from: (www.parapsych.org)

Parapsychology Foundation (2009). Organizational Description. Retrieved from: (www.parapsychology.org)

Pertzov, Y., Miller, T. D., Gorgoraptis, N., Caine, D., Schott, J. M., Butler, C., & Husain, M. (2013). Binding deficits in memory following medial temporal lobe damage in patients with voltage-gated potassium channel complex antibody-associated limbic encephalitis. *Brain: A Journal of Neurology, 136, 2474-2485.*

Psychology Dictionary (n.d.) Retrieved from: (psychologydictionary.org)

Rice University (2006). Parietal Lobe. *Language and Brain: Neurocognitive Linguistics.* Retrieved from: (www.ruf.rice.edu)

Treffert, D. A. (2009). The savant syndrome: An extraordinary condition. A synopsis: Past, present, future." *Philosophical Transactions of the Royal Socie-*

ty B: Biological Sciences 364 1351–1357.

The Free Dictionary (n.d.) Retrieved from: (www.thefreedictionary.com)

Videbeck, S. (2008). Psychiatric-Mental Health Nursing (4th Ed.). Philadelphia: Wolters Kluwers Health, Lippincott Williams & Wilkins.

Watts, A. (2009). Eastern Wisdom: Zen in the West & Meditations. The Alan Watts Foundation.

WordNet Search - 3.1 (2014). Princeton University. Retrieved from:(www.wordnetweb.princeton.edu/perl/webwn).

INDEX